BEST AVAILABLE SCIENCE

Fundamental Metrics for Evaluation of Scientific Claims

A. Alan Moghissi, Michael Swetnam,
Betty R. Love, and Sorin R. Straja

Cover Design: Patrick Worcester

Published 2010 by the Potomac Institute Press
Potomac Institute for Policy Studies
901 N. Stuart Street, Suite 200, Arlington, VA 22203
To order documents or for additional information:
www.potomacinstitute.org
Telephone: 703.525.0770; Fax: 703.525.0299
Email: webmaster@potomacinstitute.org

ISBN 978-0-615-29819-1 Paperback

TABLE OF CONTENTS

PREFACE

This book is an expanded version of the document *Best Available Science; its Evolution, Taxonomy, & Application,* first published as a monograph in 2008. It is the result of cooperation among numerous individuals and organizations. Its primary objective was, and continues to be, prescribing a taxonomy and process primarily for policymakers to use in evaluating scientific information that enhances the policy or lawmaking process. Scientists and technologists generally ascribe to the principle that good policy and legislation should be based on information and facts that are reliable and credible, using Metrics for Evaluation of Scientific Claims (MESC) or as we suggest in this book, the Best Available Science (BAS).

This book is not intended to be a prescription for, nor a commentary on, the appropriate, sufficient, or valid process for conducting science. Our understanding of the world, environment, and nature has been achieved through a variety of disciplines and endeavors. We are convinced that the taxonomy presented in this book will assist policymakers in judging the maturity of such knowledge at any one point in time. Independent peer review, reproducibility of experimental results, and objective consensus are all tools for assessing the usefulness of scientific knowledge, and can be a compass for the development of public policy and/or law.

Although not often considered in the technical literature, it is important to recognize that there are areas that are beyond the purview of science. In particular, the respective roles of religion, and societal objectives, in prioritizing various areas of science for funding and public use demand further attention, and these are specifically addressed in the present volume.

The development of the BAS concept was greatly encouraged by discussions with several distinguished individuals. These included Norman Borlaug, Nobel Prize laureate of Texas A & M University, College Station, TX and the founder of Green Revolution; Willard (Bill) F. Libby, Nobel Prize laureate of the University of California, Los Angeles; Hans Süss of the University of California, San Diego; Dixy Lee Ray, former Chairman of the Atomic Energy Commission; Frederick Seitz, former president of the National Academy of Sciences; and Melvin W. Carter, past president of the International Radiation Protection Association. Their contributions continue to stimulate the advancement of this concept despite the fact that they are no longer among us.

It is most gratifying to note the existence of individuals within the government, particularly the U.S. Department of Energy, who had the wisdom to appreciate the long-term implications of reliance upon credible and reliable science in the decision-making process. Those whose support ultimately led to the development of the BAS concept included Goetz Oertel, Clyde Frank, Gerald Boyd, and Ines Triay. Anibal Taboas also championed the cause of BAS despite the resistance of many colleagues who did not appreciate the value of the subject.

Numerous other individuals were directly or indirectly helpful in the evolution of the BAS concept. Among them were three past presidents of the American Society of Mechanical Engineers: Charles Velzy, Ernest Daman, and Nathan Hurt. Similarly, we are grateful to Richard Wilson of Harvard University for erudite discussions contributing to the concept and this project.

The authors would like to express their gratitude to Gordon Oehler and Fred Saalfeld, for reviewing the draft of the first edition of this book; they provided critical and most constructive comments, as did Dennis K. McBride, Joe Canny, Anthony Frank, and Tawfik Hamid. Kathryn Schiller and Sherry Loveless worked tirelessly on copyediting drafts of this manuscript both in penultimate and final form, and we thank Matthew Amin for his support, and Patrick Worcester for the cover design. Finally, the authors are grateful to James Giordano of the Potomac Institute Press for intellectual contribution to, and editorial oversight of this project.

INTRODUCTION

Historically, the global community has been dealing with claims and counter claims on various issues that are directly or indirectly derived from science. A claim may consist of a scientific discovery, identification of a chemical compound, a new synthesis process for a chemical compound, an invention, identification or description of a new disease, and/or assessment of risks associated with exposure to an agent or an operation, to mention a few. Many individuals—some with good will, others with more malevolent desires—have misused scientific information in an attempt to impose particular views upon various groups, or the general public, at large. Based upon this (mis)information, some advocacy organizations have claimed that certain actions taken or not taken will lead to dire, if not catastrophic, consequences. The public and the media, as well as many within the scientific community itself, can often be confused when judging the validity of these claims. In the past, it has been suggested that peer review would guarantee the "acceptability" of any scientific information. But the mere process of peer review does not afford such certainties. For example, a claim that is based upon the physical phenomenon known as *momentum*, would have a higher chance of being reliable than the prediction of the size of a fish population at a given time and given place, even though both claims are peer-reviewed.

Most disciplined science and technology professionals would agree that irrational belief, superstition, or emotionally charged reactions cannot- and should-not sustain policy. But how is a person with little or no scientific knowledge to judge the validity of information provided under the guise of science? The misdeeds of National Socialist and Communist regimes fostered broad suspicion of the ways that science can be misinterpreted, miscommunicated, and misused to suit socio-political agendas. In response, the scientific and technical community has worked hard to restore a level of trust and respect for its methodological enterprise.

As we advance the fund of knowledge and understanding, there is increasing acceptance and embrace of the validity of science over superstition, and there is a defined tendency to profess the need to favor rationality over mythic beliefs. The problem is that most science is neither absolute nor irrefutable. Those in the scientific community will acknowledge that much of what we know is constantly evolving and there are degrees of maturity to the science or knowledge we have at any point in time. Yet, advocates of one position or another often use scientific knowledge, regardless of its maturity, as if it were an absolute that cannot be refuted. Certain policymakers (including some elected officials) are frequently more influenced by public opinion polls than scientific information. This is because a policymaker can understand the meaning of a poll more than the validity (or evolving nature) of scientific information. The actions of many scientists are also based on public opinion, and one must recognize that science is broad, and a distinguished scientist in one field may not necessarily be qualified to judge the validity of claims in others. Thus, we in the scientific—including the engineering—community have a duty to provide meaningful context for scientific information, so as to frame the criteria upon which its validity might be both judged, and useful to policymakers and the public.

This book is not intended to review the historical development of various claims and counter claims. Instead, it attempts to provide a rational taxonomy and process for evaluating the validity of available scientific information that is relevant to policy and other decisions. Science inherently engages efforts to understand the nature of things, with the intent that this knowledge will guide development of policies and laws by which society functions. Therefore, the community of science upholds an obligation to assist policymakers in understanding the level of maturity of knowledge that can be- and should be- used to make policies or laws.

The desire to develop a process for evaluating acceptability of scientific claims began when the senior author of this book was employed at the U.S. Environmental Protection Agency (EPA). During the early history of the EPA, allegations of health effects of exposure to low-levels of pollutants ranged widely. For example, Gofman (1981) claimed that an overwhelming majority of cancer cases were caused by exposure to ionizing radiation resulting from fallout of nuclear weapons testing. Gofman was interviewed on numerous radio and television programs and advocated his view. After considerable debate, the National Research Council (NRC, the research arm of the National Academy of Sciences, the National Academy of Engineering, and the Institute of Medicine) formed a committee known as BEIR III (NRC 1980) that concluded that Gofman's findings were erroneous. The cause of many cancers was determined to be related to exposure to α, β, or γ radiation. Thus, we can see how a single false claim would negate the impact of smoking that is acknowledged to cause a large fraction of lung cancer, as well as (an undisputed) contribution of the genetics to the occurrence of certain cancers.

Similar allegations were made regarding cancer caused by agents emitted from chemical facilities. There were claims that there would be an epidemic of cancer unless drastic actions were taken to eliminate chemical emissions. There have been claims that pesticides are responsible for the occurrence of cancer, and it has even been postulated that replacing all man-made chemicals with "natural" materials would reduce cancer within the population.

Obviously, Gofman (as well as proponents of claims that chemical emissions are the cause of cancer) were not right. However, if one were willing to accept that they did have a point, we would only have to stop nuclear weapons testing and chemical emissions to eliminate cancer. This being the case, we could permit or even encourage unlimited smoking, and reduce or eliminate major efforts in funding for cancer research and treatment. Responding to the controversy, Dole and Peto (1981) published a paper that computed the potential impact of environmentally-caused cancer to be about 2% of the total occurrence of the disease.

As the simple example shows, given the status of claims and counter claims, it is important to develop a process for generically identifying metrics for evaluation of scientific claims (MESC). The development of the MESC process began with a literature search. Common references on science, philosophy, and related topics as described under "Evolution of Science", resulted in the conclusion that an existing process does not provide MESC or clear description of the reliability of scientific information. Equally important is that as science evolves, not all scientific information can be treated identically because while there is generally applicable and entirely reproducible scientific information (i.e.- scientific laws) there are also scientific judgments that can be modified as new information becomes available. In searching for a title, the term Best Available Science (BAS) was chosen to sort the validity of claims and counter claims. To ensure that the term was not used elsewhere in different contexts, we conducted a literature search and found neither an identification nor a definition of this term.

Although the choice of BAS proved to be reasonable, after the phrase was coined, many individuals and organizations used it without necessarily recognizing its generic applicability. In the U.S., government agencies at federal, state, and local levels often use the term "best available science" to identify information most suitable for their specific needs. In effect, they use information that ranges from scientific laws to personal opinions of experts, as long as that information is applicable to their needs. As will be shown, even though peer review is mentioned, such sources fail to describe how scientific information is to be structured to qualify as BAS as described in this book. Our choice of the term BAS as the foundation of MESC was based on the notion that:

1. *Best available* indicates a level of reliability. It suggests that there is information that is 'best', and some that is not. Consequently, it implies the need for some standard.
2. *Best science* implies that on occasion a specific area of science has not reached a reasonable level of maturity. Again here, there is an implication that one has to generically identify the level of validity of the specific scientific information.
3. Finally, *science* excludes any subject that is not science. As it will be seen later in this book, this exclusion is far from being generally acknowledged and accepted.

Although from the beginning of the development of the BAS concept we relied upon specific principles, their formulation occurred much later. Similarly, it took several years to bring the three fundamental parts of the BAS to a reasonable conclusion. As it will be seen later in this book, we have formulated four principles that constitute the foundation of BAS.

The first, *Universal Scientific Principles*, could have been just as well called methods, approaches, processes, and other principles that are used by all scientific disciplines. For example, if a study in any field of science uses statistics, the study must use statistical methods that are well known and well established.

The second principle deals with reproducibility of a scientific claim. It implies that anyone with sufficient skills, equipment and facilities must be able to reproduce the asserted claim

The third principle requires open-mindedness. The scientific community, and hopefully everyone else, must be willing to be open-minded and consider new ideas even if they contradict existing scientific knowledge.

Finally, the open-mindedness does not imply that the scientific community accepts every claim. The *Skepticism Principle* is the counter part to the *Open-mindedness Principle* and requires that the scientific community critically evaluate a new idea. We recognized the contradiction between the two latter principles (*Open-mindedness* and *Skepticism*) and hope that the BAS concept provides an approach to reconcile this contradiction. These principles led to the taxonomy of BAS resting upon three pillars or parts as follows:

1. Reliability: The reliability of scientific information consists of four categories. The first two categories (personal opinion and gray literature) are considered to be unreliable. The BAS concept considers independent peer review to be the foundation of acceptability of science. Very early during the development of the BAS concept we recognized that there are and will always be contradictory information even in properly-performed peer-reviewed literature.

Consequently, the BAS concept envisions consensus–processed science to be the approach to resolve disputes in contested areas of science. The BAS concept also provides for processes to ensure the acceptability and integrity of both the peer review and the consensus process.

2. Level of Maturity: The second part of the taxonomy of BAS deals with standardization of science in terms of its maturity. We identified four classes of scientific information (in fact three classes and one class that claims to be science but it is not). Proven science consists of scientific information that is uncontested and is principally reproducible by anyone with relevant training and appropriate equipment. The next class consists of evolving science and constitutes that area of science where the overwhelming majority of scientific advancements are made. The third class consists of borderline science, which as the title implies, includes judgment and speculation. Finally, there is erroneous information sometimes called 'junk science'.

3. Outside of Purview of Science: The third part of the taxonomy of BAS constitutes areas that are outside the purview of science. These include religious faith, societal decisions, and numerous other areas of social value in discourse. The BAS concept does not imply that the scientific community or individual scientists are without role in areas that are outside the purview of science. Instead, it explicates that the role of the scientific community (and/or individual scientist) is similar to the roles of members of any other profession. Simply- that a 'profession' dictates expertise borne of knowledge and skill in the defined, specific field, discipline, and/or area. Here, specificity is the key. We recognize the complexity of the problem, yet are convinced that the BAS concept provides a necessary pathway to resolve potential conflicts arising at the interface of science and areas of broad social and political concern.

For obvious reasons, the focus of this book is the description of the BAS process, with particular emphasis upon applications to decision processes used in policy and socio-public domains. And thus, we also believe that BAS is a logical next step in the evolution of science, and it is desirable to briefly describe the various stages of scientific development.

EVOLUTION OF SCIENCE

The efforts that eventually led to the development of BAS started with the evaluation of the classical evolution of science from its inception to the enormous advances of the last century. Science is derived from the Latin word, *scientia,* meaning knowledge. Einstein (1939) defined science as an "...endeavor to bring together by means of systematic thought the perceptible phenomena of this world into as thorough-going an association as possible. To put it boldly, it is the attempt of the posterior reconstruction of existence by the process of conceptualization." Note that Einstein was a mathematician and theoretical physicist and his definition reflects this background. There are numerous other definitions of science, including:

1. Systematic knowledge of the physical or material world.
2. Knowledge gained by systematic study.
3. A branch of knowledge or study dealing with a body of facts or truths systematically arranged and showing the operation of general laws.
4. Possession of knowledge as distinguished from ignorance or misunderstanding.

This book does not intend to review the existing literature on the evolution or philosophy of science as there is a voluminous amount of work in these areas. Instead, we attempt to find a coherent description of various steps that lead to generally applicable and accepted scientific fact(s). Most treatments of this subject describe the philosophy of science, how science evolved, etc. To be sure this is important.

Here however, the objective is to briefly describe the evolution of science so as to depict some distinction from the BAS concept described in the next section. This is not intended as a comprehensive review of the evolution of science from antiquity to today. The classical description of evolution of scientific knowledge, sometimes referred to as scientific method, is based upon the "hypothesis-theory-law" (HTL) process. Figure 1 shows the four stages of evolution of scientific information.

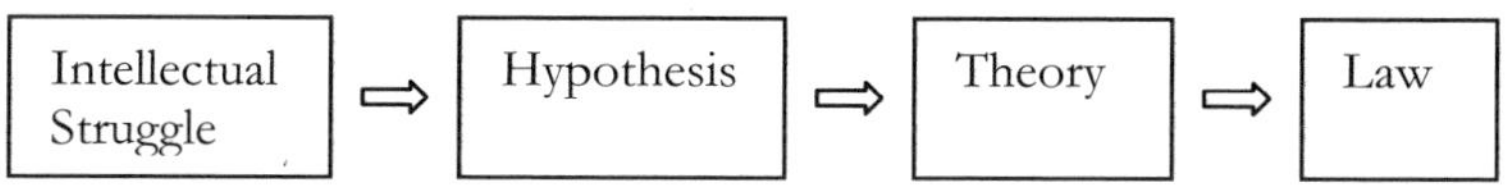

Figure1. Classical description of the evolution of science

The process begins with observation, an experiment, a thought, curiosity, or intuition. Often, this grows from skepticism of existing knowledge. Many investigators have experienced this intellectual struggle; in some cases skepticism may lead to a rationalized postulate called a hypothesis. Typically, a hypothesis is an idea whose merit must be evaluated by experiments, computations, and other empirical (or sometimes theoretical) scrutiny.

Once a hypothesis is developed, the next step is the formulation of a theory by the addition of new evidence. A theory includes accepted principles, and reasoning, but also engages assumptions that attempt to explain a wide range of circumstances. Note that there is an interaction between the verification approaches (including experimental tests) and the status of a theory. A theory is typically strengthened by validation through (experimental, and other) evidence.

Many times, the formulation of a theory is the final stage of scientific endeavor. However, the ultimate objective of the process is the conversion of a theory into a scientific law. Whereas theories include assumptions and unproven ideas, scientific laws do not. Although the scope of a law may be limited, it must be able to predict all events that applicable within its scope. Notice that scientific laws are unambiguous and describe the predictability of a scientific process.

Let us use the example of the structure of the atom to demonstrate how scientific laws have evolved and been modified:

The idea of the existence of the atom goes back to antiquarian Greece. Thanks to efforts of modern scientists such as Ernest Rutherford, Niels Bohr, Louis de Broglie, Ernst Schrödinger, Werner Heisenberg, and Max Born, among others, the structure of the atom was eventually established. It could be readily demonstrated that an atom consists of a nucleus containing neutrons and protons; the latter having a positive charge. In addition, the atom contains electrons that have negative charge equal in number to those of protons, making the net atomic charge neutral. These discoveries gave rise to various hypotheses and theories that resulted in a scientific law describing atomic structure.

Accordingly, each element has a specific number of protons and neutrons. In a stable element, the number of protons is equal to the number of electrons, and this determines the properties of that element. For example, an atom of iron has 26 protons (and electrons); if an element has 24 or 28 protons, it is no longer iron but chromium or nickel respectively. Every textbook on chemistry and physics published in the first third of the 20th century included such a description with figures that showed a nucleus surrounded by orbits where electrons would circulate.

At that time, the scientific community recognized that an atom may contain other subatomic particles, but there was general confidence that the atom was the smallest entity that retained the property of an element. It was also known that many elements have more than one isotope having the same number of protons but a different number of neutrons. On occasion, two isotopes of an element were given different names. For example, the radioactive noble gas resulting from the decay of radium was called radon, while the isotope of the same element resulting from the decay of thorium was called thoron. During the early 20th century there was an intellectual struggle that centered upon the possible occurrence of radioactivity. An event changed the status of scientific foundation of the atom when Otto Hahn and Fritz Straßman (1938) irradiated uranium with neutrons and discovered several alkaline atoms such as cesium. How could elements such as cesium—an element with 55 protons—end up in a mixture of uranium with 92 protons? Initially they were looking for impurities in chemical agents that were used but could not find anything to explain their results. As a chemist, Hahn had difficulties in identifying the source of many elements in the mixture—including cesium. Liese Meitner, his physicist colleague, explained that Hahn's experiments resulted in splitting the uranium atom, and much like a stone hitting a glass window, he had produced "glass-splinters". Did that finding change the law describing the structure of the atom? It did not; but eventually it enabled a new law describing how a large atom can be split. This new law is the foundation of nuclear energy.

In every scientific discipline there are examples of the HTL process. This process does not imply that once a law is established, no further progress is, or can be, made. To the contrary; the example of nuclear fission shows how the four processes constituting the classical HTL progression-hypothesis, identification and evaluation of the structure of atom, the existence and the nature of subatomic particles, and the modification of atomic law- have continued to this date.

TAXONOMY OF BEST AVAILABLE SCIENCE

FOUNDATION OF THE BAS CONCEPT

The BAS concept required nearly three decades of effort to develop a process that is applicable to various scientific disciplines. The concept attempts to provide metrics for evaluation of scientific claims (MESC) by answering the questions of what is science, what is the level of its maturity and its validity; and what is outside the purview of science.

During the development of the BAS we faced several issues that had to be addressed before a coherent concept could emerge. In particular, we had to reconcile the idea that a scientist must be open-minded and accept new scientific ideas, while being cautious and able to separate facts from fiction. In other words, how can one be cautious by reconciling the existing scientific knowledge with the desire to evaluate revolutionary thoughts? The result of these efforts led to the development of four core principles upon which the BAS concept is based.

1. Universal Scientific Principles

Universal Scientific Principles (USP) implies that there are certain principles and methods that are applicable to all scientific disciplines. The detailed description of USP would require significantly more and extensive efforts than would be possible within this book. Instead, only a broad outline and a few examples are provided here. All scientific disciplines use specific methods and apply the rules of statistics in sampling, analyzing, and reporting results.

2. Reproducibility Principle

Reproducibility is the true proof of the validity of a scientific claim. It separates undisputed areas of science from those that include assumptions, interpretations, and in some cases, the inclusion of ideological and societal objectives. This principle implies that any investigator who has the necessary skills, proper equipment, and the appropriate facilities should be able to reproduce the claim.

3. Open-mindedness Principle

This principle implies the willingness to accept new knowledge. Every claim of a discovery, identification of a new principle, development of a new drug, identification of a potential human health problem, and description of an environmental risk requires a willingness to carefully evaluate its assumptions and validity. The scientific community must look at all claims to avoid repeated past errors. History is full of misdeeds of governmental agencies, theocracies, and the scientific community in rejecting scientifically valid ideas because they were new, inconsistent with the existing ideology, belief, or in certain cases, prevailing knowledge.

4. Principle of Skepticism

This principle requires that those who make a scientific claim must provide sufficient evidence supporting their claim. This principle, sometimes called the "show it to me" principle, appears to contradict the *Open-mindedness Principle.* How can one be open-minded and yet skeptical? For obvious reasons there must be a balance between the *Open-mindedness Principle* and the *Principle of Skepticism.* The BAS concept provides an approach not only to achieve this balance but also to identify the level of maturity and certainty associated with the claim. A key ingredient in reaching this balance is the obligation of those who make a claim to provide sufficient evidence to satisfy not only members of their discipline, but also the scientific community at large, that their claim is credible. The USP plays a key role in achieving this balance. At a minimum, the claim must comply with USP.

EVOLUTION OF THE BAS CONCEPT

The first step in the development of BAS concept involved the evaluation of the reliability of a scientific claim. This segment of the BAS process was peer-reviewed several times (AAES 1989; ASME/RSI 2002). We recognized that in some cases, a scientific claim is based upon undisputedly valid principles. Conversely, as science continues to evolve, there are claims based on an observation with only marginal scientific validity. Once this process reached a certain level of maturity, it became clear that there are various stages in the status and maturity of science. The classification of scientific information had to be reconciled with the classical HTL process. This effort resulted in a system that included four scientific classes. As the process evolved it was brought to our attention that there were really only three scientific classes, and one class that claimed to be science, but did not qualify as science, per se.

Finally, we recognized the complexity of areas that are outside of purview of science, yet still affect the scientific decision process. This is evidenced by the misdeeds of certain regimes ranging form the interference of the church in scientific research, to dictates of various governments in the first half of 20th century. As the BAS process evolved, we observed with great interest the potential harms that can be caused by the inclusion of societal objectives in science.

This book concentrates on the description of the BAS process rather than a discussion of its evolution. A detailed description of the evolution of the BAS concept is being addressed in a subsequent volume.

Figure 2 shows the relationship between the foundation of the BAS and its components.

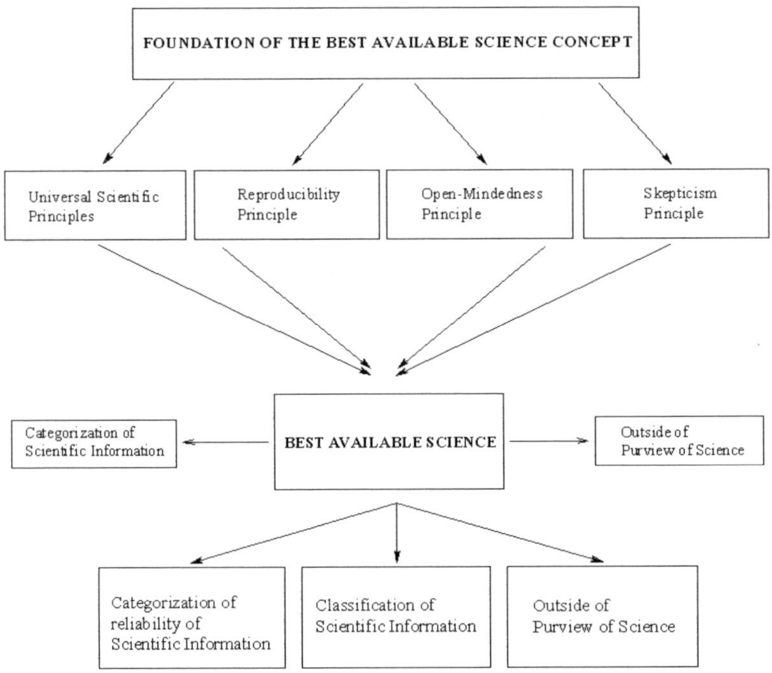

Figure 2: Taxonomy of BAS.

The concept of BAS rests on three independent, yet interrelated pillars:

1. Assessment of the reliability of scientific information.
2. Classification of scientific information based on its level of maturity.
3. Areas outside the purview of science.

1. ASSESSMENT OF THE RELIABILITY OF SCIENTIFIC INFORMATION

The value of peer review and similar processes in assessing the validity of scientific assertions has been known for at least two centuries. Mandel (1996) describes the evolution of peer review in medical research. Kronick (1991) suggests that peer review has been used since the late 18th century. It is not surprising that peer review was developed in England and expanded in the U.S. given the importance of peer consideration to the Anglo-American judicial system. This socio-judicial influence appears to have been an impetus and model for the development of the peer review process. Of note is that many European countries started to use the peer review process following its adoption by the Anglo-American systems. There are significant similarities between judicial systems' jury of peers and the scientific process of peer review.

The increasing level of financial support by various government agencies has demonstrated the urgency for the application of a fair process such as peer review in many countries of the world. Those seeking funds from government agencies rightfully ask for a fair process to distribute funds.

Based on the century-old tradition, most all-scientific publications rely upon some form of peer review in accepting a manuscript, rejecting a manuscript, or in most cases asking for revisions of the manuscript based on the comments tendered by the peer reviewers.

An area that increasingly relies upon peer review is the scientific foundation of regulations. As the impact of regulations- notably those dealing with energy, construction, food, health, and environment- has increased, society demands that their scientific foundation be evaluated to ensure that the burden imposed upon the regulated community is based upon sound principles rather than ideology; accommodation of special interests, or the arbitrary decision of the regulators. Recognizing the significance of peer review in governmental decisions in the U.S., the Office of Management and Budget (OMB) an arm of the White House, published a Bulletin (OMB 2005) mandating that various U.S. agencies develop their peer-review process following guidelines included in the OMB Bulletin. Consistent with the U.S. governmental system, the OMB published a draft proposal seeking comments from the public. During the comment period, there were numerous objections to the Bulletin from advocacy organizations, presumably because of the likelihood that peer review would reduce or eliminate such groups' use of non peer-reviewed scientific arguments in a contentious regulation. Note that the OMB provided criteria for performing peer review, the stated objective was not only to maintain a certain level of uniformity of peer review among various U.S. agencies, but also to ensure the acceptance of promulgated regulations by the regulated community. The OMB Bulletin included recognition that the level of rigor in peer review must be commensurate with the financial (as well as social) impact of various regulations.

This exemplifies the need for a formal and generally acceptable process to categorize the reliability of scientific information. In the BAS concept, scientific information is divided into distinct categories to reflect an ascending level of reliability. Figure 3 illustrates the components of reliability assessment in BAS.

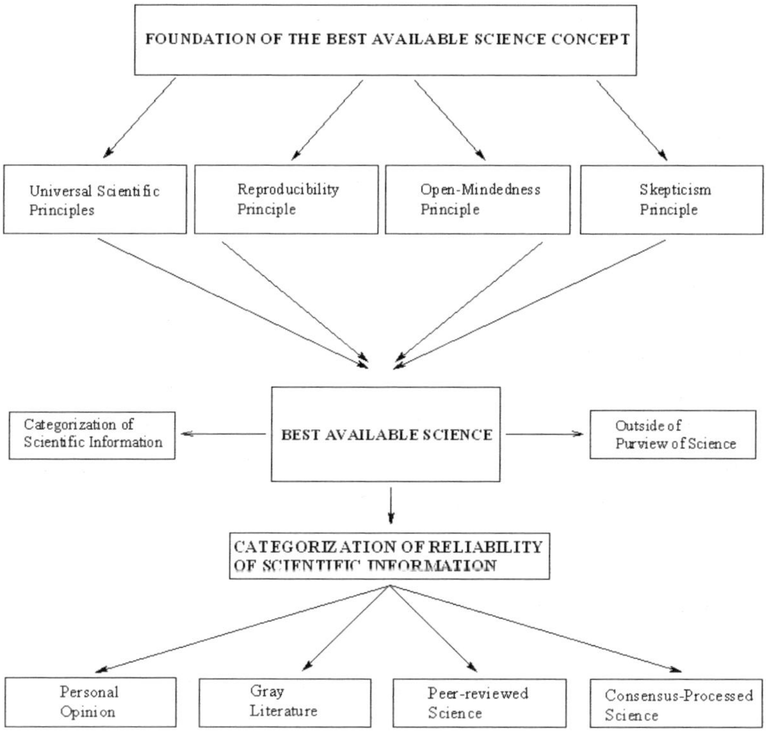

Figure 3. Categorization of scientific information in accordance with BAS.

Group 1 - Personal Opinions

The views of individuals (regardless of their training, experience, and social agenda) are included in this group. Personal opinions are seldom—if ever—BAS. At best, this category can be used to initiate scientific study of a concept, topic, area, or field. The standard process of news media often engages this information in reporting of scientific issues as such opinion is regarded as reflecting perspective(s) and/or feelings of the public, as well as particular individuals within the scientific community, as germane to the issue.

Group 2 - Gray Literature

Written information prepared by government agencies, advocacy groups, et al., that has not been subjected to an independent peer review is included in this category. This often consists of an organized and written form of personal opinions. Ideally, this category should also be used to initiate a study. The scientific quality of information included in this category can range from reproducible science to erroneous information. This category of information is often used by government agencies, advocacy groups, and individuals seeking to advance certain information (but not all relevant findings) in support of an idea, or to promulgate an agenda.

Group 3 - Peer-Reviewed Science

Information subjected to independent peer review falls within this category. By far the most reliable peer review involves confirmation by attempting to reproduce the experiment or claims. Historically, this has been the method of choice in evaluating various claims, and remains a well accepted and important form of peer review.

Probably the most famous case of reproducibility as a form of peer review was the claim of investigators (Fleischmann and Pons 1989) of having achieved fusion of deuterium (^{2}H) at slightly elevated room temperature. Due to the enormous consequence of cold fusion, several investigators quickly attempted to reproduce the experiments of Fleischmann and Pons. The results of some of these efforts were published by Gai et al. (1989) and Williams et al. (1989) and indicated flaws in experiments by Fleischmann and Pons.

The ability to reproduce a study is limited and often impractical. In disciplines such as physics, chemistry, biology, and engineering, certain studies can be reproduced, but there is not interest in doing so as this would be both cost- and time- inefficient. In other cases, such an effort is inherently impossible. Therefore, peer review is often the only mechanism to assess the validity of a scientific claim, and thus remains the foundation upon which to base the acceptability of a scientific claim. There is extensive literature describing the peer review process.

Throughout this literature, three criteria are advocated to ensure the integrity of the peer review process:

Criterion I: Qualifications of peer reviewers: A peer reviewer is an individual who is capable of performing the project, or that segment of the project that is being reviewed. The process of ensuring compliance with this criterion is demonstrated by:

1. Education in an appropriate field.
2. Sufficient professional experience in an area that is directly related to the topic under review.
3. Peer recognition as demonstrated by serving on relevant panels; election to offices in relevant professional societies; and other activities that indicate competence and involvement in the professional community.
4. Publication in peer-reviewed scientific journals, patents, and related scholarly accomplishments.

Criterion II: Independence of peer reviewers: The reviewer must be free of conflict of interest. This is one of the most complex and contested issues in peer review. Typical conflicts of interest include:

1. Having been materially involved in the project under review or having a member of the immediate family involved in the study.
2. Having financial benefit from a positive or negative review.
3. Affiliation with organizations whose activities are being reviewed, if the outcome of the review would directly or indirectly impact the income or standing of that individual.
4. A preconceived commitment to a certain scientific or engineering approach without consideration of potential alternatives.
5. Unambiguous commitment to an organization with a bias to certain scientific or engineering approaches without consideration of potential alternatives.

Criterion III: Transparency in the peer review process: A key issue is transparency as regards the process and methods of peer review. As well, such transparency applies to any organization that evaluates potential reviewers' qualifications and independence. Criteria would require:

1. Identification of the process used to select reviewers, including oversight of the process and mechanisms to ensure the enforcement of policies.
2. Description of policies regarding qualifications of the reviewers, specific to how the (four) requirements for qualification of reviewers are considered and weighed.
3. Description of policies on conflict of interest.
4. Availability of the review results, including information on reviewers' identity either directly, or by providing a list of reviewers used by a journal during the year, or reviewers used to award grants during a certain period.

Group 4 - Consensus-Processed Science

This category consists of information resulting from a process used to resolve scientific disputes, particularly in contested areas of science. The prerequisite for this process is the formation of a group of individuals that work to reach consensus on a specific scientific subject. Members of this group must meet the qualifications and independence criteria described for peer review. Furthermore, the process must be overseen by an organization that is uniquely qualified to do so. Professional societies of scientists and engineers are the primary resources for this activity. However, professional societies face certain limitations in performing this important task, including:

1. Professional societies are qualified to manage the consensus process only in their respective disciplines. For example, engineers cannot authoritatively speak on medical practice, and chemists cannot judge the validity of issues related to electrical engineering.
2. Management of the consensus process must exclude parochial interests of the profession represented by the professional society. Many professional societies represent their interests and must demonstrate that they disregard these interests during the consensus process. Often professional societies have difficulties in avoiding such interests of their members. However, if properly managed, these can be minimized or eliminated.
3. On occasion, it may become necessary to reach a consensus on issues covered by more than one professional society. This is particularly necessary for addressing multidisciplinary issues. There is ample experience indicating that professional societies are not only capable but also willing to join forces to manage the consensus process.
4. Organizations established by the U.S. Congress for the purpose of reaching scientific consensus must meet certain

requirements. For example, the National Research Council (the research arm of the National Academy of Sciences, the National Academy of Engineering, and the Institute of Medicine) is uniquely qualified to evaluate interdisciplinary scientific issues. In contrast, the National Academy of Public Administration is qualified to address administrative issues, and the National Council on Radiation Protection and Measurements is qualified to evaluate issues related to radiation. Despite the temptation to get involved in areas outside of their competence, these organizations should not and cannot perform peer review and reach scientific consensus in such areas.

2. CLASSIFICATION OF SCIENTIFIC INFORMATION

Based upon the identified need, we developed a classification for the assessment of the level of maturity of scientific information, as seen in Figure 4.

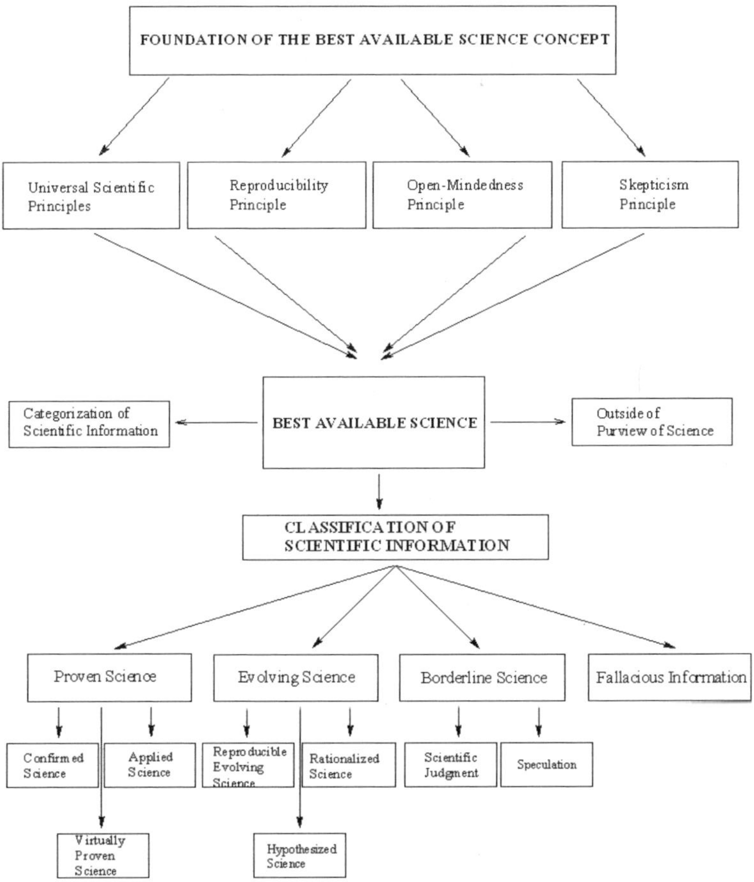

Figure 4: Classification of Scientific Information

The final outcome of our process identified the following classes:

Class I – Proven Science

This class is the equivalent to scientific laws in the classical "hypothesis-theory-law" (HTL) process. In the BAS system, it consists of scientific laws—sometimes called scientific principles—and their application. The cornerstone of this class is compliance with the *Reproducibility Principle* implying that any investigator who has the proper equipment and the necessary skills can reproduce it. Therefore, this class of scientific information does not require assumptions or any other conditions for its validity.

Class IA - Confirmed Science: Scientific laws and all other information that have been unequivocally confirmed and generally accepted constitute class IA. We recognize that each scientific law or scientific fact has its limitations and prerequisites for its proof. For example, the validity of the law of gravity has been well established, including the fact that it does not apply to the inside of an atom. Similarly, the speed of light is known and the differences in results of its measurement are within the generally-accepted accuracy.

Class IB - Applied Science: This class consists of application of scientific laws to various branches of commerce and industry. The only prerequisite for information to be placed into this class is the correct interpretation of scientific laws. Many of the engineering disciplines and a large segment of other applied sciences fall into this class.

Class IC – Virtually Proven Science: This class consists of information whose reliability has been uncontested, yet there is insufficient proof to be placed in the proven science class. The process for placing scientific information into this class consists of: 1) ensuring that reliable observation does not contradict information in this class and 2) no alternative system describes the observed phenomena. Possible examples of this class include both general and special relativity theories. These theories (or principles) attempt to describe the universe, forces operating in the universe, and their relationship. There has been evidence supporting these theories, and attempts to disprove them have failed. However, due to their nature, it is unlikely that they can be completely proven at least at present.

Class II – Evolving Science

This class does not have a counterpart in the classical HTL process. Although there is an overlap between certain elements of this class and theory of classical HTL process, this class deviates significantly from theory. Scientific advances in all disciplines represent evolving science. The current BAS concept provides only three subparts to this class, however, it is likely that as applications of the BAS concept are expanded, other subparts will be added.

Class IIA – Reproducible Evolving Science: Reliable information dealing with a subject that is not completely understood constitutes the core of this class. The key factor in placing information into this category is reproducibility. Although the observations of this class may not necessarily have general applicability, they are clearly and unambiguously reproducible by those with appropriate skills and equipment. Advancements in all branches of science (e.g. physics, chemistry, biology, etc.) are based on a desire to improve knowledge in this class.

An example is provided by much of medical science. Advancements in medical science have resulted in the need to apply methods common in other scientific disciplines and engineering to medical practice. One of the consequences of these advancements is the need for what has become known as *evidence-based medicine*, requiring a specific categorization of scientific information so as to enable appropriate use in clinical care.

Class IIB – Rationalized Science: The scientific foundation of information placed in this class is derived from proven or reproducible evolving science. However, it uses assumptions and extrapolations to derive results and conclusions. A characteristic of this class is its level of reproducibility. In order to reproduce information in this class, the investigator must have not only proper skills and the necessary equipment, but must also accept the asserted scientific foundation; assumptions, choice of mathematical processes, default data, and other prerequisites. Consequently, scientific information covered in this class does not meet the requirements of the *Reproducibility Principle.* Many predictive models and a large segment of contested scientific information fall into this class.

Class IIC – Hypothesized Science: This class consists of an organized response to an observation, idea, or any other thought initiating process. In certain respects this class is analogous to hypothesis in the classical HTL process. In contrast to the two other classes (IIA and IIB) described in this section, information in this class is not necessarily based on proven or reproducible evolving science. Although many great scientific discoveries started with this class, there is also a long list of hypothesized information that has proven to be either wrong or not worth pursuing. Let us use an example to demonstrate the point:

Imagine that an archeologist is studying an area known to have been the site of an ancient civilization. This site had been studied extensively resulting in the conclusion that it had existed around 3,000 BC. During the study the archeologist identifies an item with carbon deposits on its surface. The subsequent ^{14}C- dating indicates a time frame of about 10,000 BC. The archeologist hypothesizes a number of alternatives and informs the archeological community accordingly.

As this example indicates, the information in this class is unique and requires attention by the scientific community. However, it is imperative to recognize that any conclusion or societal decision based on this class is likely to be provocative of controversy and social reaction.

Class III – Borderline Science

As the title implies, this class is not necessarily science. The closest part in the classical HTL process to this class is hypothesis.

Class IIIA - Scientific Judgment: On occasion, information is provided to society that lacks scientific foundation. Similarly, often decisions must often be made without having the needed information, including basic principles, data, etc. The methodology for expert judgment is reasonably well developed and consists of asking a number of presumably knowledgeable individuals to give answers to specific questions and statistically assess the results. However, this often remains only slightly better than an educated guess.

Class IIIB - Speculation: This class consists of information that cannot meet standards described in any of the above classes. It is often based on the intuition of an individual who wants to stimulate a discussion or initiate a research project.

Fallacious Information

Originally, this class was considered to be the fourth class in the scientific classification system. However, during the peer review, it was brought to our attention that fallacious information is not science and therefore, does not qualify to be considered to be a class in the science category. Although this class of information clearly falls within the purview of science, it is inconsistent with the three classes identified above. Those who desire to promote specific political, ideological, or other goals disseminate some of the information in this class. Their opponents often call this information "pseudo science", "junk science", or "politically-processed science". Often the dissemination of erroneous and/or fallacious information is used to exaggerate a problem in order to move the population to accomplish a noble goal. What is being overlooked is the long-term damage that misinformation incurs.

An example of this class of science was the work of Lysenko, a Russian geneticist who claimed a new form of genetics. As a result of Lysenko's claim(s), genetics' research in the Soviet Union was profoundly and negatively affected, and this ultimately was contributory to a failure of agriscience in the USSR.

3. AREAS OUTSIDE THE PURVIEW OF SCIENCE

The BAS concept excludes areas that are outside the purview of science (see Figure 5).

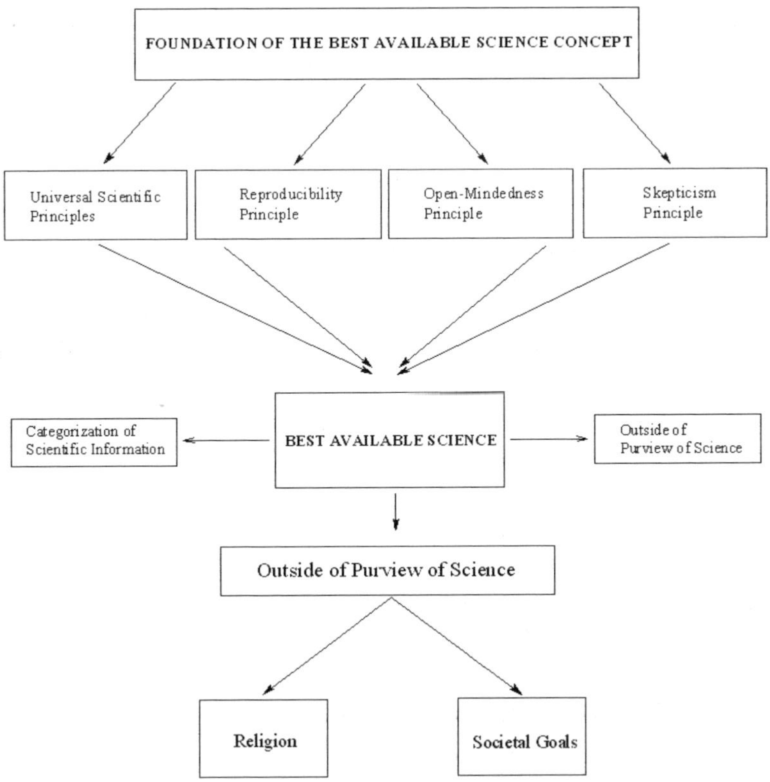

Figure 5: Areas outside the Purview Science.

Although areas outside the purview of science are in fact rather broad, for the sake of simplicity, we have addressed what we feel to be the two major domains that engage scientific information, namely religion and societal goals. Note that in some instances, societal and religious agenda may overlap, whereas in others, there may be divergence. In both cases, use (or misuse) of scientific information incurs tensions and sometimes conflicts within, and between these domains, as well as between these domains and the scientific community (NB- for an interesting overview of the distinct orientations of science and religion see: Gould SJ *Rock of Ages: Science and Religion in the Fullness of Life*. New York: Ballantine Publishing Group; 1999):

Religion

The history of religion covers millennia and is certainly not the focus of this book. To be sure, the varieties of religious belief(s), practices, convention(s), and organizations are extensive, and cannot be limited to Abrahamic traditions. Essentially, all religion(s) operate from the dictates of belief that provide dogma upon which to direct understanding and actions. In light of this, we posit that religious information can be categorized as follows:

Category I: We refer to this category as faith. Faith is based upon confidence in the truth of an idea, and entails areas beyond the address of science. One of the key characteristics is the inability of science to prove or disprove various faith-based assertions at this time, or in the foreseeable future. For example, how can science prove or disprove the existence or non-existence of God, a soul, or the interaction between God and an individual?

Category II: This category consists of ideas derived from faith (category I) to include the dogma and traditions, upon which believers rely to guide understanding and action. We refer to this category as belief. Much like faith (category I), a large fraction of beliefs cannot be proven or rejected by science. How would science judge the validity of prayer, or the value of ceremonies and practices that are representative of belief?

An example of how belief engages both scientific information and public discourse and actions is the debate and controversy surrounding abortion. One segment of population believes that a fertilized egg is human life and thus abortion constitutes infanticide. A second group claims that human life evolves during gestation and thus a fertilized egg is not yet human life. Although science can participate in the discussion, it is unlikely that it can impact the predispositions of belief in ways that allow resolution of this issue solely by the availability and use the scientific information.

Category III: In contrast to categories addressed above, there are areas where the scientific community can play a role. Information that is truly contradictory to science or proven to be incorrect is included in this category. Certainly this would include superstition. Regardless of terminology, if information in this category is detrimental to the human condition (e.g. can cause adverse effects), it is the responsibility of the scientific community to identify and describe these harms, and address and revise such information in a language that is understandable to the affected group and societal decision makers. Still, the overlap between what some consider to be superstition and others consider to be faith (category I) is such that in many cases, the result of an effort by the scientific community to intervene may be less than effective.

Societal Goals

The inclusion of ideology, faith, beliefs, societal, political, or any other non-scientific objectives in assessing the validity of scientific information is inconsistent with the BAS concept and construct(s). It is likely that while extensive, societal goals can also be categorized. Indeed, sociology, social anthropology, and political science invest considerable attention to the origin(s), nature and manifestations of societal values, schools, and use(s) of information. Given the breadth, complexity, and gravitas of sociological values and goals, we consider the categorization of societal goals to be beyond the scope of the current book.

However, it is important to note that there is ample evidence to show that the intrusion of societal objectives in the scientific process can jeopardize the objectivity- and acceptability- of scientific information. It is true that scientific investigation is performed because society, as represented by the polis and duly elected or appointed individuals or organizations, seeks to solve a problem or otherwise enhance the extant knowledge base. The scientific community is obligated to both respond to these desires in fulfilling a social trust, and to identify those areas of research, development, implementation, and funding that are more likely to succeed than their alternatives. However, it cannot be denied that scientific activities often reflect societal objectives, and may also be steered- directly or indirectly- by elected and appointed officials of government. Historical evidence including the intrusion of religious dogma to impugn the claims of Galileo, promotion of Lysenkoism, and adoption of quasi-Social Darwinism by National Socialist medicine to support eugenics, all provide clear demonstrations of the consequences of misuse and/or misdirection of societal objectives in science.

SCIENTIFIC ASSESSMENT

Previous chapters of this book have not adequately addressed the process for performing scientific assessments. Although the number of peer reviews, including independent peer reviews, far exceeds the number of scientific assessments (including their independent versions), this latter category plays a key role in a variety of societal decisions. Let us first describe the nature of each category:

Independent Peer Review

As described previously, independent peer review consists of a critical evaluation of a study, a manuscript, scientific foundation of a regulation, program; grant submissions, report making scientific claims covered under gray literature, or any other document that includes scientific information. Typically, at least three "peers" are asked to peer review the document. The qualifications of the reviewers, including their independence, the transparency of the process, and other requirements were described in previous chapters of this book.

Independent Scientific Assessment

Independent scientific assessment is the critical evaluation of a topic. In contrast to peer reviews, during the scientific assessment the existing scientific materials are collected and critically evaluated. The outcome of a scientific assessment is a document that provides the reader with the status of science addressing a specific topic. The process requires that consensus is reached on often-contradictory information in the literature. Scientific assessment is particularly useful in addressing evolving science and scientific judgment.

Objectives of Scientific Assessment

Often an industrial organization, academia, or a government agency must deal with an issue. The objective of (independent) scientific assessment is to respond to the scientific need of the sponsoring organization asking for help. Accordingly, scientific assessment consists of a critical evaluation of a subject including the following:

- Assisting the sponsoring organization in resolving complex issues.
- Advising the organization of an appropriate course of action or pathway to reach the stated goal.
- Assisting the sponsoring agency in deciding alternative courses of action.
- Assessing existing scientific information on a specific topic and providing the sponsoring agency with the results of the assessment.

The Process

The process of identification of individuals who will be asked to participate in the scientific assessment is identical to the process used to identify individuals who participate in peer review. Much like peer reviewers, members of an Assessment Panel (AP) must meet the requirements of peer review by being qualified and independent, implying that they have no conflict of interest. Independent scientific assessment uses the process described under consensus-processed science.

For example, let's pose that a government agency needs to know what to do about regulating a toxic agent that has widespread usage and provides an economic value to a segment of the society. The questions that the agency would need to address include:

- What are the true impacts of the toxicity of that agent in various segments of the population?
- What are the economic costs of treating the toxic effects of the agent?
- What would be the economic impact of banning the agent in the affected community and other communities that are indirectly benefiting from the application of that agent?
- If the agency decides to permit limited use of the agent, what would be the most expedient approach to communicate the required cautionary information to the affected community?

These questions imply that the panel performing an independent scientific assessment must search the literature, critically review the collected information, and reconcile any/all contradictory information. The outcome of such an effort would be a report that responds to each question. If the anticipated outcome of the assessment is a scientific answer, the work of the AP is complete once the answer is provided. However, in the majority of cases, the agency would prefer to take action that would likely be outside the purview of science. In such cases, the BAS process would encourage the AP to identify alternatives for possible courses of action, and would describe potential consequences for each alternative, but the AP would not be qualified to specifically recommend a course of action.

There are several organizations that perform scientific assessments, in the U.S., these include professional scientific societies that singularly, or in cooperation with other professional groups, engage this task. In addition, the National Research Council (NRC) has traditionally provided scientific assessments for various government agencies addressing a wide range of topics. Examples of topics include: agricultural research at the National Institute for Occupational Safety and Health (NRC 2008a); conflict of interest in medicine (NRC 2008b); corrosion (NRC 2009); and ballistic imaging (NRC 2008c).

Ideally, an independent scientific assessment is based entirely on documents whose reliability cannot be questioned. Examples of these studies are:

1. Peer-reviewed information classified as proven science or reproducible evolving science. If necessary, peer-reviewed information included in rationalized science can also be judiciously included.
2. Laws, regulations, and other legal information.
3. Information such as reliable monitoring data.

In practice many independent scientific assessments include gray literature, rationalized science, or even borderline science. In these cases, the outcome of the study would have to be peer-reviewed.

Recognizing that the NRC must rely upon information that does not always meets the requirements identified above, the NRC has established a peer review process. The NRC has a Report Review Committee that is formed to oversee the review of NRC reports. In practice, this effort is delegated to the commission or other entity that has appointed the group responsible for writing the report. The peer review is performed by a group of at least three individuals who receive several questions (review criteria) and prepare answers to these criteria. An individual is chosen to act as the overseer of the process in a manner similar to the editor of a scientific journal who evaluates the responses and the revision of the report.

BAS VS. BEST AVAILABLE INFORMATION

As the BAS concept evolved, we observed that numerous terms were being used (by a variety of groups) in an attempt to identify information that was suitable for describing and/or defining what elements and practices constitute (or might be valid for) "good science". However, in the overwhelming majority of cases, the objective was to define or identify what types of information could meet a specific need, rather than defining the quality of the science itself. For example, the Endangered Species Act addressed "best available data", and required regulators to search and identify available data, using their judgment to discern what information best sustained an impending decision. Similarly, the Clean Water Act addressed "best available technology", by mandating that the Administrator of the U. S. Environmental Protection Agency (EPA) would identify that technology best suited for application to water problems. In describing its strategic plan, EPA states "…[the] EPA uses the best available scientific approaches, data, and models to anticipate potential threats, evaluate risks, identify solutions, and develop standards that protect the environment and safeguard human health." Clearly there are many terms that refer to the identification of desirable information, particularly in the legislative process. An example of the perception of "best available science" by the U. S. Congress is the Energy Policy Act of 2005 that suggests inclusion of peer-reviewed studies.

Another example of using the term "best available science" while actually applying conventional scientific processes is the law passed by the State of Washington (Sheldon et al. 2005), that mandates that all cities and counties within the State designate critical areas—including wetlands—within their boundaries and protect them. The document details an ideal publication that could be useful for regulating wetlands.

It also identifies the process of managing wetlands when information is less than desirable or useful. Consistent with legal requirements, a number of communities in the State of Washington have published a version of "best available science".

The American Fisheries Society (Sullivan et al. 2006) defines "best available science" for its implementation in fisheries and identifies an ideal study that would be applicable to fisheries. The document provides a definition of science and how each part of a paper should be organized. Accordingly, elements of a high-quality scientific paper consist of:

1. A clear statement of objectives;
2. A conceptual model, that is the framework for characterizing systems, making predictions, and testing hypotheses;
3. Good experimental design and a standardized method for collecting data;
4. Statistical rigor and sound logic for analyses and interpretation(s);
5. Clear documentation of methods, results, and conclusions; and
6. Peer review.

As this description shows, there is a general inconsistency in definitions of "best available science", "best available data", "best available technology" and "sound science". Consistent with the BAS concept, it thus becomes necessary to identify and define a collective term that clearly and unambiguously describes the intention of these terms. It is likely that Best Available Information (BAI) covers all processes that describe technical information applicable to a need, provided that an attempt is made to identify and sort all available information. Ideally, during the identification of BAI, a process is used that is analogous to the criteria described under assessment of reliability of scientific information.

Also of interest is how science can lead to the development and manufacture of a new product. In the context of this book, a product consists of anything that is produced or manufactured, such as a new drug, new car, new aircraft, agricultural product, or a kitchen machine. Figure 6 depicts the process.

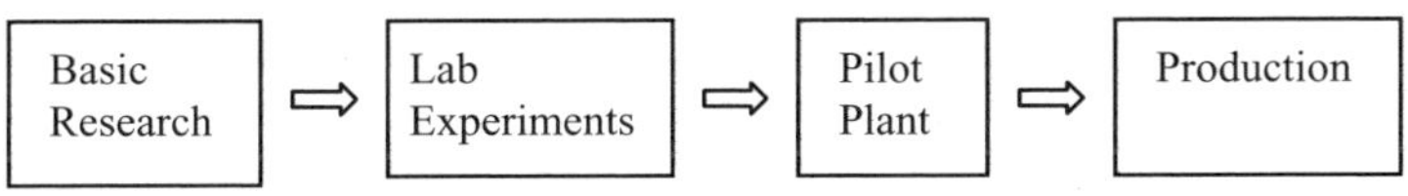

Figure 6. The production process

Typically, development of a product begins with basic research to seek knowledge. In certain cases, this provides an opportunity to evaluate the potential for developing a new product. Once potential usefulness is elucidated, an attempt is made to develop an evidentiary outcome or product in the laboratory. In practice, the validity of claims in basic research and development of outcomes and/or products in the laboratory are confirmed by independent peer review. The next step in the process is small-scale or prototype production referred to as pilot plant activity. For industrial-scale production facilities, the idea is to scale-up the laboratory operation- that have typically been performed under controlled conditions- to a larger scale where the process is inherently less controllable. Similarly, for production of items such as an airplane or a car, one or more items are produced and tested to ensure the validity of laboratory experiments. For production of drugs, testing for efficacy and safety is included in this third stage of the process.

Once the pilot stage has been completed, a full-scale operation is performed. Here again, the process is inherently less controllable than in the pilot stage. The latter two stages of the process are often sufficiently unique so that those who are intimately involved in it can judge the level of success. In addition, in a free-market society, commercial interests of the producers can - and often do - prohibit release of information to outsiders, including independent peer reviewers.

This description should not be construed so as to infer that there is little value or significance in processes that are collectively identified as BAI. Many processes included under BAI utilize independent peer review- a significant requirement of the BAS concept. On occasion, commercial or national security interests may mitigate independent peer review. Conversely, BAI is sometimes used as an equivalent or replacement for BAS where neither a commercial/ national security interest would preclude the application of BAS. In short, BAI is valuable- and valid- in those circumstances and instances when/where BAS requirements may not be allowed, feasible, or practical.

APPLICATION OF BAS

Society in general and the scientific community in particular, frequently must decide the acceptability of countless claims. Whereas some of these claims are entirely or predominantly scientific, there are assertions dealing with numerous societal issues with a clear or hidden scientific foundation. The BAS concept provides metrics for evaluation of scientific claims (MESC). The BAS/MESC mitigates consideration of societal objectives, determines the reliability of scientific information, and then evaluates its level of maturity by placing it in the appropriate class of scientific information. It may be recalled that the BAS concept was based on four principles:

1. Universal Scientific Principles
2. Reproducibility Principle
3. Principle of Open-mindedness
4. Principle of Skepticism

These principles led to a process with three independent and yet related parts, sometimes called footings or foundations:

1. Evaluation of reliability of scientific information by placing the information into one of the four categories: personal opinion; gray literature; peer-reviewed; and consensus processed.
2. Classification of the information into four classes: proven science; evolving science; borderline science; and fallacious information. Note that the first three classes include more than one subset.
3. Recognition that there are areas outside the purview of science.

A number of scientific publications follow the BAS concept. Most scientific journals have a formal peer review process. Furthermore, in areas of evolving science, every journal has provisions for submitting letters to the editor correcting potential errors, amplifying shortcomings of a claim, or otherwise identifying potential applications of a scientific issue.

Probabilistic vs. Deterministic Scientific Information

During review of the previous version of this book, questions were raised as to the status of science that requires the inclusion of probabilistic processes. A key question was: Can a probabilistic process qualify as proven science?

Let us use an example to address the subject: In the U.S. and many other countries, charitable, religious, and community organizations use raffles to raise funds to support their goals. Let's assume that 25 individuals participate in the raffle and each participant pays $10. Let us also assume that there will be one prize-winner, and the prize is $150. If the process is fair, each participant has one chance in 25 to be the prize-winner. The rule of probability cannot determine which one of the participants will win the prize. However, the probability that one of the participants will win is 100%. If a scientific claim would follow a process that is analogous to this example and otherwise could be classified in one of the three BAS classes, the inclusion of the probability would not change the assignment to the appropriate class. In other words, the probabilistic process does not influence assigning a scientific assertion to a BAS class.

Application of Universal Scientific Principles

As previously stated, there is general agreement within the scientific community that universal scientific principles (USP) can be applied to all scientific disciplines. We have shown the need to use appropriate statistical methods, unbiased sampling, application of relevant analytical methods, and reproducibility of results. Based upon the requirements for assessing the reliability of scientific information in common practice, a scientific study must be reviewed by those who are peers to the author of the study (i.e. peer-review). A closer look reveals that in the overwhelming majority of cases, any such study consists of two distinct parts:

1. Unique features of the discipline upon which the study is based.
2. Application of the USP to derive specific results and conclusions.

A key feature of the BAS process is a separation of these parts. Whereas certain features of each discipline remain unique, USP are used by all disciplines. Consequently, the acceptability of scientific claims (by any discipline) depends upon 1) acceptance by members of that discipline (for discipline-specific science) and 2) members of other disciplines (for USP).

There are significant advantages in applying the BAS concept to highly contested scientific areas, including the reduction or elimination of potential parochial interests of a particular discipline. In most contested areas of science, members of a discipline tend to avoid making judgments that would restrict or negate resources that are important to continued activities of the discipline (or field). The inclusion of individuals outside a discipline when evaluating the proper use of USP both enlarges the pool of potential reviewers that are free of conflict of interest and ensures fair and frank assessments of methods, process, and outcomes, so as to reduce, if not remove specific bias(es).

The implications of applying BAS, as described in this section, are more significant than they may first appear. If a scientific claim proves to be valid, the inclusion of members of other disciplines who are capable of assessing the proper application of USP will undoubtedly enhance the chance of general acceptance by the larger scientific community and the general public.

There are well-established processes that rely upon both USP and unique requirements of a discipline. Note that these processes included both open-mindedness and compliance with the *Skepticism Principle*. A summary of these processes is as follows:

I. Reliance upon Existing Science: The investigators provide evidence that based on one or more proven scientific principles or the combination of proven scientific principles. In this case, they must describe these principles and provide sufficient data to demonstrate that the underlying assumptions that relate to the applicability of these principles are not flawed. The overwhelming majority of scientific and engineering projects supported by federal agencies, as well as those supported by private industry, fall into this category. In effect, the investigator uses new approaches and applies them to existing scientific principles.

II. Reliance upon Acceptable Data: On occasion, an investigator claims to have developed a new system with an unknown scientific foundation or a new scientific principle. In many cases the underlying scientific principles do exist, but are unknown to the investigator. However, on rare occasions, there have been observations with truly unknown scientific reasoning. This latter situation was widespread in past centuries but is increasingly uncommon. However, the scientific community cannot rule-out such a possibility, and, based on historical evidence, is well equipped to evaluate the validity of these assertions. Among the numerous criteria, the following four are essential and are commonly used in evaluating these assertions:

1. Reproducibility by the Investigator: The investigator must be able to unequivocally reproduce the claimed information. For example, field measurements require demonstrated reproducibility of the background, as well as the measurements upon which the claims are based.

2. Verification by Other Methods: The investigator must be able to compare the results obtained from using the new system with data that were independently obtained. Using the above example, the investigator must also demonstrate that the system can measure the same parameters with the accuracy as measured by other systems.

3. Verification by Other Investigators: The system developed by the investigator must be verifiable by uninvolved investigators. Reproducibility is one of the cornerstones of science and is a key to the acceptability of scientific claims. Again, referring to the example above, others must be able to reproduce measurement results using the system developed by the investigator.

4. Consistency with Existing Science: The investigator must prove that claims are not inconsistent with proven science (class I). In other words, the reported results do not contradict the extant, accepted laws of nature, although it has been shown that there are limitations to the applicability of certain scientific laws. In such cases, the investigator must have overwhelming and convincing evidence demonstrating that there is no other choice but to accept the establishment of a new scientific law or the modification of an existing law.

III. Application of USP in Peer Review: A key issue application of USP is the role in peer review. The two categories of information described in this section require independent peer review to become acceptable. A prevalent and significant misunderstanding of peer review persists relating to contested areas of science. Recall that scientific conclusions are based on two distinct areas of any scientific discipline:1) discipline-specific topics, and 2) application of USP to that specific discipline to reach the claimed conclusion.

In light of this, individuals who are qualified in a discipline are thus qualified to review discipline-specific topics. In contrast, qualified individuals who are outside that discipline can, and should, review the application of USP regardless of the topic.

ASSESSMENT OF MODELS AND CORRELATIONS

Every scientific discipline uses models to evaluate a process, predict future events, and/or describe a potential cause and effect relationships. Unfortunately, the reliability of predictive models is often inconsistent: while prediction by certain models is highly reliable, others provide little more than an educated guess. In the following discussion, an attempt is made to determine the level of scientific maturity of models and correlations, and categorize them in appropriate scientific classes of scientific information.

Correlation vs. Causation

Do storks bring babies? There is evidence suggesting a statistical correlation between the size of a stork population and the number of births (Höfer et al. 2004). Similarly, there have been a number of allegations that a particular person- or groups of people- incur social harms. One need only recall witch-hunts and pogroms to find examples of such ideas and their effects. What both of these situations reveal is the meaning(s) and critical distinction between correlation and causation, and the importance of this distinction to models. To initiate this discussion, let us examine how the traditional scientific "hypothesis-theory-law" (HTL) process has addressed the problem.

Based on principles derived from physical sciences, it was common to require that an observed correlation should meet two criteria in order to be acceptable:

1. The correlation must be statistically significant.
2. The observed effect must be scientifically plausible.

In response to ambiguities fostered by these two criteria, Sir Austin Bradford Hill (1965), a British epidemiologist, subsequently expanded the criteria necessary to relate correlation to causation; these were:

1. **Strength in association:** The statistical association must be sufficiently greater than a few percent.
2. **Temporality:** Exposure must precede effect.
3. **Consistency:** There must be multiple observations gained in different situations.
4. **Theoretical plausibility:** The observed correlation must be scientifically plausible.
5. **Coherence:** The association must be aligned with other knowledge.
6. **Specificity in the causes:** The effect should have distinct and discriminable cause.
7. **Dose-response relationship:** Alterations in the dose (i.e.- amount) of one factor must affect another, as consistent with the observed association.
8. **Experimental evidence:** Experiments should support the correlation and causation.
9. **Analogy:** Accepted phenomena in one area can be relevant and applicable to other areas.

In recent years, these so-called Hill criteria have been optimized as follows:

1. **Temporality:** effects are seen if- and only if- preceded by effect.
2. **Strength:** There must be a reasonably strong statistical association.
3. **Dose-response:** The dose-effect relationship can be direct or inverse but must be consistent with the observed association.

4. **Consistency:** There must be repeated observations by different investigators, in different places, different circumstances, and different times.
5. **Coherence:** Cause and effect relationships must be congruent with current knowledge of the natural history of the event(s) involved.

Despite efforts to correlate statistical association with causation, the confusion continues as demonstrated by repeated allegations that an adverse effect was 'caused' by an event, chemical, or an activity. Note that, as stated above, the issues related to correlation vs. causation are, in fact, a subset of models that predict the occurrence of an event (including adverse effects). Consequently, it is reasonable to discuss these in the context of predictive models. Let us use the BAS concept to address these issues. The BAS concept divides predictive models into several groups based on their expected reliability.

Predictive Models

A large number of policy decisions dealing with budgets, economics, environment, safety, transportation, human health, and ecology are based on predictive models. Given the large economic consequences of model application, it is surprising that in many cases, highly reliable and reproducible models receive nearly the same acceptance as uncertain or speculative models. The BAS concept provides metrics to evaluate scientific claims (MESC) and thus a mechanism to—at least qualitatively—assess the reliability of predictive models.

Predictions Based on Proven Science: Models that are entirely based on proven science (class I) are, by definition, applied science. For example, models that predict the fall of an item on the surface of the earth are normally based on the law of gravity. Consequently, they are precisely predictable and the time required for the item to fall can be accurately computed. Similarly, most chemical reactions fall into this category. The reaction between silver nitrate and sodium chloride always leads to a white precipitate consisting of silver chloride. The reaction occurs in any part of the world as:

$$AgNO_3 + NaCl \rightarrow AgCl + NaNO_3$$

The entire field of engineering, medical instrumentation, space science, and numerous other parts of industrial production use models that are based on proven science.

Predictions Based on Reproducible Evolving Science: As the title implies, these models rely entirely upon reproducible evolving science. The distinction between these models and models based on proven science is that the scientific foundation of the former is not completely understood. An example of these models is the relationship between smoking and lung cancer that involves the exposure to and carcinogenic effects of high-levels of condensed aromatic compounds (IARC 1998).

Primary Predictive Models: Although the foundation of a large number of models used in the contested areas of science is proven science (class I) or much of the reproducible evolving science (class IIA), they also use assumptions that fall into neither class. Therefore, their predictions include inherent uncertainties. Consequently, primary predictive models are entirely rationalized science (class IIB). Many models that comply with Hill criteria (or their revised version) fall into this group. It is imperative to recognize inherent uncertainties associated with these models.

For a number of reasons, society may have no other choice but to use these models in making decisions. However, as will be discussed in this book in the section dealing with ethical requirements of BAS, experience shows that once science advances, a primary model can be elevated to models based on reproducible evolving science. Models may also prove to be wrong. The example of smoking and health demonstrates the ascension from primary predictive models to models based on reproducible evolving science. Conversely, the so-called Shaken Baby Syndrome (SBS) has moved in the opposite direction. Initially, it was assumed that shaking a baby with sufficient force causes subdural hematoma and thus the SBS. Studies by Uscinski and McBride (2008) demonstrated that the scientific foundation of SBS was flawed. Consequently, SBS falls into the category of erroneous information within the BAS taxonomy.

Secondary Models: These use primary models as their foundation, including those that rely somewhat or predominantly upon rationalized science (class IIB). The predictive ability of these models is significantly lower than those of primary models. These models are likely to fall into borderline science (class IIIA, scientific judgment). A large number of models attempting to correlate two parameters without considering revised Hill criteria fall into this group. Whereas the principle of open-mindedness dictates that these models be considered for further study, a society that bases its decisions on these models must consider the possibility that such models are flawed.

Tertiary and Lower Models: These models use secondary models as their foundation. The predictive ability of these models is at best consistent with borderline science (class IIIB, speculation). The predictive ability of tertiary models is so poor that they are at best borderline science (class IIIB, speculation) or more likely erroneous information. A society that bases its decisions on these models must accept the notion that it may well be wasting its resources.

Reliability of Peer-Reviewed Information

The role of peer review (and consensus process) is to ensure compliance with the Principle of Skepticism and the Open-mindedness Principle. However, there are two issues of concern in assessing the reliability of the results of independent peer review or independent consensus process:

I. Inherent Shortcomings of Peer Review: Much like any other human endeavor, the peer review process is not perfect. As stated in the introduction of this book, it is likely that the Anglo-American judicial process necessitating jury of/by peers was used as an impetus for peer review in scientific publications. In modern times, both systems (jury of peers and peer review) evolved in England and their use was expanded in the United States.

Let us use the analogy of the judicial process using trial by a jury of peers, to assess the level of perfection of the peer review system. In the past, when an individual was convicted by a jury of peers, in order to reverse the conviction, the individual had to demonstrate that the jury made a mistake. Typically the convict had to find the real culprit; show that he/she was somewhere else when the crime occurred; or provide other evidence that often was difficult to produce. Despite the difficulties, occasionally it was possible to reverse the conviction of the accused. However, the advancement of technology has provided an avenue to re-evaluate the previous convictions. Parker et al. (2003) reviewed the case of death row inmates who later were found to be innocent. In most cases, the advancement of methods such as DNA typing played a key role in the final decision. According to Parker et al. (2003), during the period from 1973 to 2002, a total of 107 death row inmates were exonerated. Florida, with 23 cases, led the number of exonerations, followed by Illinois, with 17 cases. Clearly in the absence of available advanced technologies, most of these inmates would have been put to death.

There are those who suggest that due to its lack of perfection, peer review should be abolished or reduced to a *pro forma* process. What choices are there to replace evaluation of the scientific validity of an assertion? Is the answer reliance upon government employees who may or may not have scientific qualifications; or advocacy organizations that clearly are trying to promote their preconceived ideas? Who else is there? Are scientific decisions of these or any other individual or group of individuals likely to be perfect? What is being overlooked is that peer review increases the probability that the Skepticism Principle is met and thus increases the chance of confirmation of the validity of a scientific claim.

II. Shortcomings of the Process: There are numerous reasons for the imperfection of the peer review process including the following:

1. The subject is so specific that the number of potential reviewers is limited, or most reviewers have a conflict of interest.
2. The required qualification of the reviewers is insufficiently determined.
3. Qualified reviewers are unavailable if a specific deadline must be met.
4. The review criteria are too general and the reviewers go on a "fishing expedition".
5. No provisions have been made to separate societal objectives from science.
6. There is a lack of recognition of the significance of the process.
7. Preconceived opinions of the reviewers impact the results of the review.

Many of these shortcomings can be reduced or avoided by forming an oversight group as described under consensus-processed science in this book. However, despite best efforts, experience shows that errors do happen. What is the solution?

III. Inherent Uncertainties of Science: As described in the section on taxonomy of science, there are several classes of scientific information with different levels of maturity. In the case of class I (proven science), notably class IB (applied science), there is a probability that peer review results in highly reliable information. Similarly, there is the reasonably high likelihood that peer review of information in class IIA (reproducible evolving science) will have similar results.

Starting with rationalized science (class IIB), even if the peer review process were perfect, the results will have inherent uncertainties. Consequently, the reliability of peer review is at best equal to the reliability of the science in that class. The information in classes II and III have lower reliabilities and thus the results of peer review can, at best, identify the level of reliability.

Comparison of BAS with HTL

It may be recalled that the classical description of evolution of scientific knowledge, sometimes referred to as the scientific method, is based on the so-called "hypothesis-theory-law" (HTL) process. The process starts with several potential events collectively referred to as intellectual struggle. This struggle may lead to an organized system called hypothesis. Subsequently, the hypothesis is converted to a theory by the addition of new evidence. Typically, a theory includes assumptions, accepted principles, and other reasoning that attempts to explain a wide range of circumstances. Finally, a theory is converted into a scientific law. Whereas theories include assumptions and unproven parts, scientific laws lack both of these factors.

As stated above, the BAS concept determines not only the reliability of the information but also assigns it to a specific class. It is imperative to appreciate that science evolves and the accepted knowledge may require modification, amplifications or, on occasion, revision. Consequently, skepticism, critical thinking, and lack of blind acceptance of the results of an experiment or computations of a model are key ingredients of science. The following examples demonstrate the need for skepticism and compare these two systems:

The Atom: In discussing the HTL process, the example of evolution of the atom, including nuclear fission, was described. Modern atomic theory is in fact proven science (class IA). There are those who argued that an atom is not the smallest component of the nuclear system. However, a closer look indicates that the atom was defined as the smallest version of a matter that retained the properties of a specific element. For example, an atom with 26 protons and the appropriate number of neutrons is iron (^{26}Fe). If the nucleus of the element number 26 (iron) is changed by removing some of the protons, that atom is no longer iron. To be sure, most elements have multiple numbers of neutrons resulting in isotopes of that element. Traditionally, isotopes of each element are shown with the number of protons and the total number of protons and neutrons. Let us use the example of iron to demonstrate the point. There are a number of iron isotopes including ^{59}Fe, ^{60}Fe, ^{61}Fe, etc.

Once it was shown that the nucleus of an atom (in this case ^{235}U) could be split, it was repeatedly reproduced; the rationale and the process were identified; and their application was demonstrated. Thus, nuclear fission is placed in the class I.

This example demonstrates that the BAS concept and HTL process yield identical results. (Note that on occasion the subject described in this section is called atomic theory. We leave the terminology of the HTL process to their proponents.)

Phlogiston: The concept of phlogiston was proposed by Johann Joachim Becher in 1667. Becher modified the antiquarian concept of the four natural elements (fire, water, air, and earth) and replaced fire and air with phlogiston. According to the phlogiston theory, flammable materials contain phlogiston—a substance without mass, odor, taste, or color.

According to Becher's phlogiston theory, upon combustion of inflammable materials, phlogiston is released. Consequently, the dephlogistonated materials could no longer be burned. There were several other investigators, including Jan Baptista Van Helmont and Georg Stahl, who were able to reproduce studies by Becher.

Based upon reproducibility it appeared that the phlogiston concept was valid. According to the HTL process, phlogiston was considered to be a theory. Similarly, according to the BAS concept, phlogiston would have been classified as reproducible evolving science (class IIA).

However, Lavoisier showed that oxygen in air was responsible for the heat generation resulting from combustion of flammable materials. He demonstrated that in the absence of air (containing oxygen), inflammable materials would not burn and in this way, negated the idea of phlogiston as the flammable substrate. There is a considerable literature detailing the rich history of the phlogiston theory (Higgins 1789) and the contributions of Lavoisier (Bell 2005).

Stomach Ulcer: Historically, it was believed that peptic ulcer was caused by dietary, stress, and certain behavioral factors. Accordingly, treatments focused on bed rest, consumption of certain foods, and general remedies. As described by Munnangi and Sonnenberg (1997), gastric acid was considered to be the cause of ulcer. Consequently, antacid medication and drugs that blocked acid production were prescribed. The medical literature showed peer-reviewed information indicating beneficial effects of these remedies. In fact, data showed that 75% of ulcer patients were treated with antisecretory medications. Based on this information, the causation of ulcer, as described above, would fall into the reproducible evolving science (class IIA) or at least in the rationalized science (class IIB).

A study by Warren and Marshall (1984) conclusively demonstrated that *Helicobacter Pylori* was overwhelmingly the cause of peptic ulcer, a conclusion that was confirmed in a workshop organized by the (U. S.) National Institutes of Health (NIH 1994). The new information demonstrated that the previous science was flawed, indicating that information that is placed in the reproducible evolving science can indeed be wrong.

BAS Requirements for Multidisciplinary Projects

Multidisciplinary projects or subjects are of particular interest in contested areas of science. Consequently, the formation and operation of a review panel (RP) or consensus panel (CP) require procedures to ensure the integrity of these processes. As stated in the section addressing Universal Scientific Principles (USP), there are two reasonably independent parts in any scientific study: discipline-specific and USP. Each discipline included in a multidisciplinary project or subject requires that at least three qualified individuals participate as peers in the process. In addition, at least three individuals must be qualified as peers in USP. As a general rule, many individuals who are qualified in discipline-specific areas are also qualified as peers in USP. The reason is that a peer in a discipline-specific area must have used USP to qualify. Consequently, a reasonable structure of RP or CP would consist of the following:

1. A panel (RP or CP) is formed consisting of at least three individuals for each discipline. In practice, the number of disciplines would have to be limited to ensure that the process remains practical.

2. The three members (or more if necessary) of each discipline review their respective parts and prepare their findings and recommendations.
3. The outcome of all discipline-specific panel deliberations is then presented to the entire CP or RP to reach conclusions on the subject.
4. The participation of those who are not qualified in USP in the deliberations of the CP or RP would be limited to ensuring that the integrity of the process is maintained.

Practices inconsistent with BAS: As expected, there are numerous examples of processes that are inconsistent with the requirements of BAS. The two most common examples are:

1. An organization—scientific, governmental, or industrial—forms a panel to evaluate a specific subject or a project. It has been common practice to include "a representative" of a discipline or a group in a panel that evaluates, reviews, or comments on a subject or a project. Let us assume that there are legal issues involved in this process. In this case, it would be insufficient to have one individual representing the discipline of law. The BAS concept would require that at least three individuals who are peers in the specific area of law and are independent (i.e.- lack of conflict of interest) participate in the process. In this case, the three individuals reach an agreement on responses to the criteria (questions) provided to them. If they are a part of a lager panel, the role of the panel is to ensure that the three reviewers follow the instructions and otherwise follow the pre-agreed upon process.

2. The media, political leaders, and advocacy organizations often claim that there is "a consensus" on a specific subject.

Consensus does not imply that the majority of published papers follow a conclusion; or an organization (or even most organizations) agree about a trend; or most prominent scientists agree about a subject. According to the BAS concept, the consensus results from a formal process. In this case an organization with sufficient credentials—preferably without parochial interests—would have to establish a panel consisting of peers who are independent (lack of conflict of interest). This panel is provided with specific criteria that exclude societal objectives; it prepares findings responding to each criterion; and it makes recommendations that are derived from the findings. Those who make claims without going through such a process are unfamiliar with scientific processes in general and the BAS concept in particular.

Responsibilities of Scientific Organizations

Professional societies can and should play a key role in providing members of their respective professions, the scientific community, the media, and ultimately the general public consensus-processed scientific information. Every medical specialty in the U.S. has developed guidelines resulting from a process similar to the process described in the BAS as consensus-processed science. There is a significant distinction between guidelines covering scientific and related issues, and those dealing with professional ethics.

Scientific Issues of the Profession: As described previously in this book, each profession is uniquely qualified to address areas covered by that profession. In doing so, the profession must consider not only discipline-specific science but also the USP. The organization representing the profession would provide a service not only to the members of its profession but also to the society as a whole by developing guidelines related to the scientific activities of its members.

It is imperative that in doing so, the organization avoids addressing parochial interests of the profession.

Ethical Issues of the Profession: Many professions have developed ethical guidelines addressing the practice of its members. Whereas some of these guidelines are derived from legal requirements, others represent the views of some or many members of the profession. It is important to recognize that guidelines reflect and are derived-at least in part- from societal values that are a) contemporary and b) both affect and arise from members of the profession. As a discipline, ethics is concerned with the systematic analysis and enactment of those decisions that have moral importance. Thus, while there is indubitably ethics *in* science, and an ethics *of* science, there is ongoing debate about whether ethics *is* a science, or whether it falls within the purview of science.

However, as Giordano (2006, 2010) and Giordano and Boswell (2009) have recently noted, what remains indisputable is that any ethical analysis of scientific conduct (in research or practical application) must be based upon the most contemporary factual information. This 1) provides ethical reasons for BAS, 2) further strengthens the need of BAS to any meaningful ethics of science, and 3) sustains BAS as vital to upholding science in the public trust.

ETHICAL REQUIREMENTS OF BAS

Consistent with the BAS concept, society as a whole has developed ethical requirements for the scientific community. Virtually every professional society in the U.S. (and many other countries) has developed a code of conduct, and whereas some deal with professional behavior, others address societal issues related to the specific discipline. There are several organizations that engage scientific assessments; in the U.S. these include professional societies that work either singularly, or in cooperation with other groups in discussing relevant scientific information. There are numerous ethical issues that arise in scientific research and practices. A thorough discussion of these is beyond the scope of this book. However, one of the most important ethical issues in science is the treatment of human subjects in research (and by extension, clinical care). Current ethical guidelines for the treatment of human research subjects are based upon the Nuremberg Code and subsequent Belmont Report.

Organizations engaged in human subjects' research are required to establish and maintain an Institutional Review Board (IRB) that is responsible for reviewing all proposed experiments that include human subjects. A key to the approval of a proposed experiment is 1) that the subject must be fully informed of any and all potential benefits, burdens, and harms of the procedures involved, and 2) that the subject would benefit from the proposed experiment (DHHS 2005). Consistent with the BAS concept, legislative and executive branches of government have imposed these requirements.

In many ways, the use of BAS reflects the ethical mandate to use information soundly as bespoken by Hippocratic Oath of the fourth century BC. The Hippocratic Oath (NIH 2009) has been modified numerous times and the most recent version as adopted by the American Medical Association in 2001 (AMA 2001) includes the following:

"We the members of the world community of physicians, solemnly commit ourselves to:

1. Respect human life and dignity of every individual.
7. Educate the public and policy about present and future threats to the health of humanity,
9. Teach and mentor those who follow us for they are the future of our caring profession.

The American Chemical Society (ACS) has had a code of conduct since 1965. The most recent version of the code of conduct of ACS (2007) states:

> "Chemical professionals have a responsibility to serve the public interest and safety and to further advance the knowledge of science. They should actively be concerned with the health and safety of co-workers, consumers, and the community. Public comments on scientific matters should be made with care and accuracy, without unsubstantiated, exaggerated and premature statements."

The code of Ethics of the American Institute of Chemical Engineers (AICHE 2003) includes the statement that AICHE members shall "... issue statements or present information only in an objective and truthful manner."

As these examples indicate, honesty and truthfulness are important to codes of conduct of all major scientific professions. However, certain advocacy organizations and special interest groups depend upon and often use scientific information that is, at best, primary or lower group models or in some cases frankly erroneous information. In addition, individuals and organizations may make claims about a scientific issue not because of a potential gain but rather because of lack of knowledge. An important and less-than-fully recognized obligation of the scientific community is to provide recipients of scientific information (e.g.- scientific peers, members of other scientific disciplines, and the general public) with information that expressly includes:

1. The reliability of the claim.
2. The level of maturity of scientific information including any associated uncertainties; assumptions; and default data.
3. The scientific community is limited to the assessment of scientific claims, and excludes consideration of those issues that are outside the purview of science.

It is equally important to maintain an appropriate level of skepticism. Clearly, an individual scientist can—with confidence—claim the validity of proven science including its three classes IA, IB, and IC. Obviously, the information must be reliable and at least category III (peer-reviewed). A minimum level of skepticism is required for reproducible, evolving science (class IIA). As discussed later in this book, although it is unlikely that such information will be proven to be incorrect, historically this has happened, and therefore, its occurrence in the future cannot be ruled-out. Skepticism is necessary when regarding any rationalized science (i.e. class IIB) and the level of skepticism increases with subsequent classes (III A and particularly III B). It is true that some scientists may have a "hunch" about the direction of future outcomes, or the likelihood of *secondary models* to be valid. However, such hunches fall into the category of belief and are, by definition, outside the purview of science.

As previously stated, scientists do not have a monopoly on deciding what is good for society. Consequently, once science is evaluated using the peer review or consensus processes, members of other professions (e.g. lawyers, accountants, or publishers) are equally qualified to decide what is good for societal aims and ends. The scientific community must abide the ethics that are derived from end reflect the moral values and codes of society, and scientific organizations are encouraged to develop codes of ethics that obligate reliability and maturity of any and all data used in pronouncements that are disseminated to the communities served.

APPLICATION OF BAS CONCEPT TO VARIOUS DISCIPLINES

The Introduction to the chapter on Taxonomy of BAS suggested that the writings of Shakespeare, Molière, or Schiller could constitute knowledge, as could music of Beethoven, Mozart, and Verdi. But would such knowledge be considered science? Recall also that during the evolution of the BAS concept, we struggled with classification of various disciplines (ranging from physics to sociology) and concluded that such a classification would be unnecessary if the BAS concept were applied to all fields.

The application of the BAS concept to various disciplines has been the focus of this book. At this point, it is important to ask, what would be the role of science in non-scientific disciplines? In particular, how would (or could) the BAS concept be applied to these disciplines? The following examples demonstrate applications of the BAS concept to various disciplines:

Law

The law is derived from actions of the legislative, regulatory, and judicial branches of government. Let us address each part of the legal process separately. The scientific community has a role in the law by providing the necessary scientific information to the legislative branch of government. This information could include the reliability of the science under consideration, and its level of maturity. Once this information is provided to legislators, subsequent decisions are entirely outside the purview of science.

Similarly judicial decisions occasionally call upon the scientific community to provide information that is relevant to the legal process. However, judicial decisions consist of interpretation of laws and thus are also beyond the scope of science. The situation is somewhat different for regulatory processes, as a segment of regulations includes science and thus requires the participation of the scientific community to inform certain decisions. The promulgation and implementation of regulations, however, remain beyond the purview of science. Thus it may be concluded that the practice of law requires access to various scientific disciplines, but, law is not a scientific discipline.

Mathematics

The oft heard phrase "math and science" colloquially implies that mathematics is not a part of science. What is overlooked is that mathematics was the foundation of scientific advancement in both the classical Greek and Arabic cultures. According to the BAS concept, mathematics qualifies not only as science, but often as class I - proven science. Arithmetic computations meet the *Reproducibility Principle*, as do analytical geometry, algebraic equations, and many other domains of mathematics.

Medicine

Clearly modern medicine is a scientific enterprise, although the practice of medicine entails other, non-scientific (e.g. humanitarian) values and activities. Historically, medicine- as both a discipline and practice- 'evolved' from non-scientific social rituals and beliefs. However, since the turn of the 19th century, medicine has been uncontrovertibly reliant upon science, technology, and the information provided by their intersection. This trend is reflected by the conjoinment of national agencies that have provided administrative guidance and oversight to medical research and practice. The National

Academy of Sciences was established by President Abraham Lincoln in 1863 as an advisory organization to the agencies of the U.S. government. The National Research Council (NRC) was established in 1916, primarily to review proposals seeking government funding; a task that later was performed by the National Institutes of Health, National Science Foundation, and numerous other agencies. Each of these (scientific and technical) agencies have been engaged in funding and directing medical research, and (albeit at times indirectly) clinical therapeutics. However, by the mid- 1960s, the breadth and pace of scientific and technological progress germane to medicine was such, that the necessity for a federal agency specifically dedicated to medical research and practice, became evident. This organization, the Institute of Medicine, was established in 1970 as part of the National Academies According to the BAS concept, medicine is a scientific discipline. Scientific medical information is broad and covers virtually all classes of BAS. Consequently, medicine- as a field- constitutes a major application of science and its significance is recognized by a sizable financial commitment of many governmental agencies.

Music

So we finally come to the question of whether the compositions of Beethoven, Mozart, or Verdi qualify as science. Helmholtz (1863) addressed the perception of tone and described how musical sound can be improved and enhanced through application of scientific constructs of physics. To be sure, scientific principles can be used to study music: musicology, musical harmony, and instrumentation would typically be placed somewhere between reproducible evolving science and rationalized science. In this way, music is a scientific discipline; however, products resulting from that discipline are exactly that- products. Thus, according to the BAS concept, the compositions of Beethoven, Mozart, and Verdi are not science, per se, but rather products of science, and these products can be scientifically studied.

APPLICATION OF BAS CONCEPT TO NEWS MEDIA

The BAS concept was specifically developed to respond to societal needs to assure the validity of scientific claims. As described in this book, the BAS process includes:

1. The reliability of information assured by using independent peer review and independent consensus processes.
2. Underlying information classified as proven, evolving, borderline, or fallacious.
3. The limited and defined purview of science.

It is also important to recall that there were four categories of information identified for science: personal opinion, gray literature, peer-reviewed, and consensus-processed.

Dictionaries describe "news" as being "noun plural but singular in construction", implying that it is the plural of "new". Based on this definition, the task of the news media is to report whatever is new. There are three well-identified parts to news:

1. A decision on what event or issue to cover.
2. A factual description of the event or issue.
3. Interpretation of and commentary on the event or issue.

Currently, many reporters cover all of these aspects. Although the application of the BAS concept (the four principles, the four categories of reliability, and the four classes of information) to the news media would be of considerable interest, a complete discussion is extensive and is beyond the scope of this book. Instead, this chapter addresses reporting of science and scientific issues in the media.

Until the end of the 19th century, many individuals were fundamentally capable of comprehending an entire scientific discipline. Therefore the press could, with confidence, quote a recognized scientist on a specific scientific issue. The task of the reporter was only to find the correct specialist. Individuals such as Ernest Rutherford, Marie Curie, Robert Koch, and many others could authoritatively express an opinion in their respective disciplines. The exponential expansion of science and the introduction of interdisciplinary fields have made it difficult if not impossible for any individual to be knowledgeable in the entirety of a given discipline. Despite this, the press continues to quote individual scientists, based upon his/ her association with a recognized institution or university; or having received an award such as the Nobel Prize. What is overlooked is that a Nobel-Prize winner may be knowledgeable in a specific domain of a field, but not its entirety.

Much like the analogy between a jury of peers and peer review, the news media appear to occasionally follow the judicial approach that attempts to have witnesses both for the prosecution and for the defense. In reporting news that is partially or entirely scientific, the media identify individuals with opposing views. Having faced criticism that the "opposite of science is ignorance", on occasion news agencies and reporters will try to find an individual who "represents" science.

In reporting science or scientific issues, the media often use phrases such as "research shows", "science indicates", "there is a consensus", "many scientists agree", "science has decided", etc., to indicate a particular viewpoint as espoused by certain potential sources of information. It must be recognized that this rarely represents a complete description or presentation of the relevant scientific information.

Without doubt, news media face an extremely difficult problem. General journalism training typically does not include advanced coursework in the sciences or scientific methods. To be sure, there are specific courses in scientific journalism, and there are some science (including engineering) graduates who decide to enter the journalism profession. Such individuals bring a competence to reporting scientific information that is important, necessary, and laudable. However, far too frequently, scientific issues are reported by journalists with little or no (meaningful) scientific training. In such cases, miscommunication or even frank error and misuse of scientific information (and sometimes nonscientific information taken to be scientific) is possible if not probable. To illustrate, let us use the example of an individual who has some training in biology. How could such an individual judge the validity of claims in astronomy or nuclear safety? It is unlikely that an individual— regardless of training, education, and intelligence, would be able to judge the claims of the entire realm of scientific disciplines. What is a journalist to do in trying to sort scientific facts from fiction? How can someone with limited scientific education decide that one scientific claim is likely to be correct while others are not?

These problems have been aggravated by the emergence of the Internet. Since the invention of the printing press to the appearance of newspapers, there has been some attempt to control the quality of news. This control was maintained during the early days of electronic media (i.e.- radio, and subsequently television). But, with the advent of 24-hour TV news stations, the need to have broadcast material led to a gradual loosening of such control. (That is not to say that all print, TV, and radio news is cavalier. Surely, this is not the case; but with the general laxity of control and an increasing market-ethos, the news consumer is advised to be cautious in discerning information) The appearance of the Internet has allowed posting of any type of information.

Still, there are ways to access and acquire credible scientific information. Such resources include:

1. Online sites that post peer-reviewed scientific information
2. Media that maintain long-standing reputation for fair and accurate reporting
3. Juried or expert refereed news' sites
4. Reports issued by or through reputable scientific agencies

How might BAS be applied to media forums? As an example let's imagine a professional or trade organization of the media, a newspaper, a TV station, or any other entity that conjoins a group of individuals who have only basic scientific knowledge but are thoroughly familiar with the BAS including its MESC process. For the sake of simplicity, we shall refer to this group as BAS Assessment Panel (BASAP). This group would ask each reporter to provide information to the BASAP to make a judgment in the following areas:

1. Was the science derived from peer-reviewed information?
2. Based on classification of science, in what class with this information be placed?
3. Did the information include areas outside the purview of science? That is, did the information include societal, religious, or other judgments that are, by definition, non-scientific?

Note the lack of similarity between BASAP and the "science court" proposed by Kantrowitz (1961). The science court was intended to evaluate the scientific validity of claims. As described above, it is impossible to find individuals who can judge the validity of claims in every scientific discipline. In contrast to science court, BASAP would follow a process that would provide the public with an assurance of the reliability and level of maturity of the information reported. It would also minimize or remove any potential bias of the reporter (whether intended or unintentional).

BAS-BASED NATIONAL AND INTERNATIONAL DECISIONS

Numerous governmental, scholarly, and professional organizations (collectively referred to as decision-making organizations) are engaged in the development of regulations, reports, pronouncements, and other publications that include science. Many decisions are initiated by advocacy. The objective of advocacy is to promote an idea or a cause that is clear to an individual, group of individuals, nation, or group of nations. In many cases, advocates try to pass a law, promulgate regulation, initiate the preparation of a report, or force an action to accomplish their preconceived goal. Because the objective is to achieve the goal of advocacy, consistency with scientific principles may be of little concern. Therefore, advocacy often is not consistent with scientific principles. However, regardless of the desires of advocates, decision makers must ensure consistency with scientific principles to guide the use of scientific information and any/all decisions that rely upon, employ, or affect science.

Organizations that are involved in decisions that are based on science include (at least in part):

1. National (federal), state (provincial), and local governmental agencies.
2. National scholarly organizations such as National Academy of Sciences (and three organizations formed under its auspices: National Academy of Engineering, Institute of Medicine, and National Research Council); National Academy of Public Administration; National Council on Radiation Protection and Measurements and their counterparts in other countries.
3. Professional societies of scientists, engineers (and their international organizations).

4. Various scientific organizations formed under the auspices of the United Nations including the World Health Organization; the former United Nations Scientific Committee on Effects of Atomic Radiation; United Nations Educational, Scientific, and Cultural, Organization.
5. Private scholarly organizations.

Note that the majority of scientific information results from research and development performed at universities, and/or national and international research institutions. The contribution of these organizations to the BAS-based activities consists of publication of scientific papers in peer-reviewed media. These may not be correctly or completely used in advocacy.

Thus, the simplest method to demonstrate consistency with scientific principles is to apply the BAS concept to the scientific content of any advocacy. It is important to recall that the BAS concept deals entirely with science and expressly excludes areas outside the purview of science. Decisions that rely upon proven science (class I) are always consistent with the BAS concept. To some extent, the same is true for decisions that fall into the category of reproducible evolving science (class IIA). However, all other decisions require participation by the scientific community and compliance with requirements of the BAS concept.

Let us consider the case of information that falls into rationalized science (class IIB). The process starts with searching the scientific literature with the objective to compile information in a coherent report. This includes evaluation of the reliability of each reference. By far the most important criterion is independent peer review. In many cases, the individual responsible for the literature search identifies materials that fall into categories II and III (gray literature and peer-reviewed). The inclusion of gray literature in the process must be based on the judgment of whether or not it contains information that, if it had been peer-reviewed, would make significant contribution to the decision process.

As rationalized science (class IIB) inherently includes assumptions, the most reasonable approach is the formation of an independent consensus panel. Members of the panel must meet the requirements of independence (lack of conflict of interest) and relevant qualifications. The panel should evaluate the validity of these assumptions not only as being reasonable, but also to ensure that they exclude societal biasing. The assessment of both issues (reasonable intent and content and exclusion of societal bias) does not imply that the consensus process should not consider the utility of the outcomes of the consensus itself, or ultimately the decisions such consensus may inform. In fact, a properly formed panel must ensure that the results of the process provide information that meets the needs of the decision makers as much as possible. Another key issue in the formation of a panel is to include panel members who are versed in USP. Discipline-specific panel members may be inclined to compromise applications of USP to their respective discipline. Adding reviewers who are versed in USP to the panel significantly reduces this resulting problem. In summary, the outcome of an optimally-managed consensus process should be:

1. A thorough review of the existing information: This requires primary reliance on peer-reviewed information. If there is information that has not been peer-reviewed but its inclusion would significantly impact the results of the process, then this information must be subjected to independent peer review outside of the consensus process. Note that the consensus panel may or may not be qualified to review such information.
2. Descriptions of the scientific foundation of potential options available to the decision maker.

3. Proper application of the USP to include identification and description of uncertainties associated with each option.
4. Categorization of each model (primary, secondary, etc.) if modeling is to be used in the studies.
5. Acquisition and provision of any other scientific information that would assist a decision maker.

This process is applicable to other areas of scientifically relevant decision-making, as described in this book. It is often claimed that such a process is too involved and too expensive for any organization other than well-funded governmental agencies. However, this is not the case; the scientific community, at least in the U.S., has been repeatedly willing to donate time and effort in the public's interest.

As noted elsewhere in this book, the BAS concept separates science from its societal (including religious) implications. Many government agencies must make decisions based on rationalized science (class II B) or even borderline science (class II). In these cases, using the outcome of consensus processes (or any other process that evaluates scientific information), the decision maker can-, and should, describe the science, its level of maturity/ uncertainty, and thus justify any decisions made.

Finally, a brief criticism of the current practices may be in order. Decision makers often fear public scrutiny and seek input from the scientific community to provide particular information that could help to support a position, cause, or constituency. There have been allegations that decision makers often choose individuals to participate in peer review (and other scientific activities) in order to receive scientific endorsement of preconceived decisions. While, in the short term, this may expedite decisions and accommodate specific advocacy, historical evidence shows that long-term consequences of directing or misusing scientific resources can be dire.

CASE STUDIES

Before case-based examples are provided, it should be noted that the authors have diverse views on societal implications of contested areas of science. Therefore, the inclusion of an example should not be construed as an endorsement or rejection of any specific societal objective. The following examples were chosen arbitrarily and the reader will undoubtedly identify others:

Universal Scientific Principles (USP)

Love Canal: The uncontrolled disposal of chemical waste was of particular concern during the late 1970s, and eventually led to passage of the Comprehensive Environmental Response and Liability Act (CERCLA). That law mandated the establishment of a large fund, that became known as "Superfund". One of the highly publicized cases was the so-called Love Canal, an uncontrolled disposal facility not far from Niagara Falls in Upper New York State. Vienna and Polan (1984) studied potential adverse health effects in the population living near the site. They studied children born between 1940 and 1953 when the site was operating, and 1954 to 1978 when no further dumping took place. The results of the study showed that children born near the site during the operation of Love Canal had a lower weight as compared to those born near a control site. This epidemiological study consisted of two parts:

1. The organization of the study, and the evaluation of various potential health effects, that were the domain of the discipline of epidemiology.
2. The statistical evaluation of the data that fall into the domain of USP.

Whereas epidemiologists should review the first part, those who are not epidemiologists but are familiar with USP can also evaluate the second.

Dental Amalgam: There have been arguments attesting the potential toxicity, particularly in children, of mercury used in dental amalgam. Bellinger et al. (2006) published an extensive study that concluded that there were no significant differences between children who had dental amalgam and those who did not. The study consisted of two parts:

1. The assessment of neuropsychological and renal effects-within the domain of the discipline of the authors.
2. The statistical evaluation of the data and other aspects of the study that are in the category of USP.

BAS and the Ethical Obligations of Scientists

BAS is vital to fulfill the ethical obligations of science to uphold public trust and contribute to governmental protection of the polis. This is most notably upheld by testimony of scientists (including engineers and members of the medical profession) before the legislative, regulatory, or judicial branches of government as relates to the assessment and descriptions of current scientific information. The following examples illustrate this point:

Asbestos litigation: Asbestos is a term used for a number of minerals that have a tendency to separate into fibers. It is a commercial rather than a mineralogical term. Various species of asbestos have been used in buildings and other industries for insulation and as a fire retardant.

Inhalation of asbestos fibers has been linked to several adverse health effects. For obvious reasons, asbestos fibers have to become airborne to cause diseases. Inhalation of airborne fibers of certain asbestos species may cause specific pathology, including:

1. Mesothelioma, a cancer of the lining of the chest and the abdominal cavity.
2. Asbestosis, fibrous scarring of the lung.
3. Increased risk of smoking-related lung cancer.

In the U. S., both the EPA and the Consumer Products Safety Commission have banned the inclusion of certain species of asbestos in a number of products. However, products in older residential buildings and industrial plants may still contained asbestos and its removal is associated with the associated occupational exposure. In addition, there are several individuals who were exposed to asbestos prior to its regulatory restriction. Consequently, the U. S. Court system has addressed and continues to address claims of asbestos exposure.

The number of legal claims related to asbestos is exceedingly large, and the courts must assess a variety of medical/scientific issues went addressing these claims. Papers have been published that describe asbestos-related diseases and their diagnoses, and these have it was sedated the risk(s) of potential false positives and false negatives in diagnoses. A review of these papers is beyond the scope of this book, and interested readers are referred to a comprehensive description of the subject including the standard by the International Labor Organization, and classification of various occupational lung diseases, By the American Thoracic Society (2004). The National Institute for Occupational Safety and Health, of the Centers for Disease Control and Prevention (CDC) has developed occupational standards for qualifying physicians to evaluate x-rays depicting alleged asbestosis. These physicians are referred to as B-readers.

Recognizing potential income, the suits seeking remuneration for asbestos exposure mushroomed. According to a judgment by Janice Graham Jack, a federal judge in Texas (ABA 2007), there were 40 plaintiffs filing suits in Mississippi in 2000; about 76 in 2001; and about 19,672 cases in 2002. There were fraudulent cases, such as an individual who filed two claims using the same x-rays under different names. As well, there were allegations that some B-readers evaluated numerous x-ray films within a few hours, and that B-readers were paid more if the radiograph was determined to be positive.

A landmark study by Gitlin et al. (2004) demonstrated the problem. About 500 chest radiographs obtained from plaintiffs' attorneys were read by B-readers hired by the attorneys. Six consultants, also B-readers, were asked to evaluate the same radiographs but were not informed that these were obtained from plaintiffs. More than one B-reader/consultant read the same radiograph. The results demonstrated the unethical behavior of certain B-readers: whereas 95.9% of plaintiffs' B-readers rated radiographs as positive, only 4.5% of consultants did so.

The code of ethics for B-readers requires not only honesty in professional activities but also that "... B-readers shall recognize the limitations of chest radiograph classifications, and shall not make clinical diagnosis about the pneumoconiosis based chest radiograph classification alone" (CDC 2008).

The BAS concept is likely to place the reading of chest radiographs in rationalized science (class IIB). If only one B-reader evaluates the radiograph and provides a written report, the results would qualify as gray literature (category II). Thus we posit that the lawyers representing the plaintiffs violated three requirements of BAS:

1. They should have used at least three B-readers to evaluate the chest radiograph.
2. They should have given the B-readers more than one radiograph without telling them which one came from plaintiffs.
3. They should have been paid B-readers the same amount for each set of radiographs.

United States Congress

The United States Congress is expected to make decisions that are entirely consistent with basic scientific principles. There are, however, examples to the contrary.

Solid Waste Disposal Act: Axiomatically, one would expect that the Solid Waste Disposal Act (SWDA) would deal with solid waste. Instead, section 1004 (27) states "the term 'solid waste' means any garbage, refuse, sludge, from a waste processing plant, water supply treatment plant, or air pollution control and other discarded materials, including solid, liquid, semisolid or contained gaseous materials…." The law dates back to 1965, but has been amended numerous times since that time.

BAS Assessment: Obviously the U. S. Congress did not recognize the fact that 'solid' is a scientifically defined term. As used in the SWDA, the definition is inconsistent with the BAS concept.

Supreme Court of the United States

Although it is of considerable interest to evaluate decisions of the Supreme Court and their consistency with the BAS concept, a complete compendium and BAS- based analysis of Supreme Court decisions is beyond the scope of this book. Therefore, we have selected what we believe to be three paradigmatic cases for illustration:

Daubert v. Merrell Dow Pharmaceuticals: This case dealt with allegations that Bendectin, a prescription drug marketed by Merrell Dow Pharmaceuticals Inc., responsible for birth defects in two children following mothers' prenatal ingestion of the drug. The petitioners had argued their case in the district court and then in the Ninth Circuit U.S. Court of Appeals. Apparently the Supreme Court recognized the generic significance of the case and chose to consider it. The significance of the case was also recognized both in the scientific/medical and the legal communities as demonstrated by the rather large number of *amici cureae* filed by various organizations.

Amici curiae filed for the petitioners included Attorney Generals of the states of Idaho, Montana, Texas, and South Dakota; and the Association of Trial Lawyers. *Amici curiae* filed on behalf of the defendant included a number of professional societies including the American Association for the Advancement of Science, America Medical Association, and American College of Legal Medicine; various trade organizations including the Chamber of Commerce and Pharmaceutical Manufacturers Association; scholarly organizations such as the Carnegie Commission, Defense Research Institute; and the New England Journal of Medicine.

Both sides were able to provide expert witnesses of high professional caliber. The witness for Merrell Dow provided several peer-reviewed publications demonstrating that Bendectin was not a teratogen. Witnesses for the petitioners claimed that *in vitro* and (certain) animal studies showed a potential for teratogenicity of Bendectin. In addition, they claimed to have re-analyzed peer-reviewed studies including those that were already provided by the defense, and showed that birth defects were indeed associated with Bendectin.

A description of legal arguments—notably those related to the validity of the so-called Frye rule versus "Federal Rules of Evidence"—is too broad to be fully addressed here. However, the issue of scientific acceptability of evidence provided to the court is certainly reflective of the Supreme Court Decision to rely upon the BAS concept.

In making any decision, a Supreme Court judge who is facing the acceptability of a testimony "must make a preliminary assessment whether the testimony's underlying reasoning or methodology is scientifically valid and properly can be applied to the facts at issue.

Many considerations will bear on the inquiry, including whether the theory or technique in question can be (and has been) tested, whether it has been subjected to peer review and publication". According to the Court, "Scientific methodology today is based on generating hypotheses and testing them to see if they can be falsified. Indeed, this methodology is what distinguishes science from other fields of human inquiry."

The court correctly described the evolution of science (i.e. scientific method) from hypothesis to verification (or in theory). In applying a BAS-based analysis to this decision, it is important to note that the witness for Merrell Dow relied entirely upon several peer-reviewed studies that confirmed that Bendectin is not a teratogen. According to the BAS concept, this evidence qualifies as reproducible evolving science (class IIA).

BAS Assessment: The witness for petitioners provided two sets of evidence. The first consisted of a re- analysis of previously published information that was not peer-reviewed. According to the BAS concept that information is categorized as gray literature. The witness for petitioners also provided *in-vitro* and animal studies that showed the potential teratogenicity of Bendectin. Because it would take significant assumptions to convert those data to humans, that evidence is at best rationalized science (class IIB) or scientific judgment (class IIA). The court correctly relied upon reproducible evolving science rather than gray literature or rationalized science. Thus, the court's decision was consistent with the BAS concept.

Kumho Tire v. Carmichael: This case dealt with an accident that occurred when the right tire of a minivan driven by the petitioner blew out. In the accident, one of the passengers died, and others were severely injured. The tire was steel-belted and had been driven for sometime. The expert witness for the petitioner claimed that a defect in design or manufacture caused the blow-out. The defendant disputed the claim of the expert witness and provided evidence that the methodology did not meet criteria established by *Daubert.* During the appeals process, the Eleventh Circuit Court noted that "the Supreme Court in *Daubert* explicitly limited its holding to cover only the 'scientific context". The Supreme Court disagreed and suggested that there is "no relevant distinction between 'scientific' knowledge and 'technical' or 'other specialized' knowledge".

BAS Assessment: The decision of the court provided legal evidence that requirements included in *Daubert* are equally applicable to engineering and other relevant areas. As stated in the description of the *Daubert* case, the court's decision dealing with *Kumho* was consistent with requirements of the BAS concept.

EPILOGUE

This book describes a comprehensive process to evaluate scientific claims. It is based on four fundamental principles: *Universal Scientific Principles* (USP), *Reproducibility Principle*, *Open-mindedness Principle*, and *Skepticism Principle*. The BAS concept provides metrics for evaluation of scientific claims (MESC), places these claims in a clearly identifiable class, and ensures that science remains free of non-scientific criteria and objectives. A key feature of the BAS concept is the identification of USP consisting of methods, approaches, and techniques common to all scientific disciplines. Consequently, separating discipline-specific science from the USP makes it possible to evaluate the latter by the scientific community, thus improving the acceptability of a scientific claim.

The authors of this book are convinced that the BAS concept and resulting MESC have a wide-range of utility. Initially we focused our efforts on application of the BAS concept to the evaluation of scientific claims in legislative, regulatory, and judicial decision processes known as regulatory science. Several examples included in this book demonstrate the employment of the BAS concept in these decisions.

Another important application of the BAS concept is in the evaluation of scientific claims by advocacy organizations. The overwhelming majority of these organizations use science to promote their causes. The BAS concept also has applicability to news media. One needs to recognize the insurmountable problems that news media face in reporting scientific issues. There is evidence that in many cases, public perception is inconsistent with the science upon which the issue is based. The BAS concept provides methods that news media can use to improve communication of scientific information to the public. We are convinced that the BAS concept and resulting MESC (with appropriate modifications) can be applied even more broadly.

The BAS concept can be applied to projects that require new and/or unique technologies. The project may be the construction of a new airplane, new power plant, new vehicle for space exploration, or a new technology for defense activities. Let us assume that all of these cases involve several individual segments or parts that must be evaluated. The BAS concept and MESC provide an approach to subdivide the entire project into various sections and proceed as follows:

1. For each step, the required science and technology is identified.
2. The level of maturity of the science and technology identified is assessed.
3. The reliability of science and technology is determined.
4. During these three steps, societal objectives are identified and excluded. For example, most advanced technologies involve commercial and political, interests that can affect the relative "value" of technical information; these must be addressed and extricated in order to enable objectivity in the decision process.
5. If necessary, a consensus panel is appointed to resolve potential problems, reconcile potential differences, and reach a consonant decision.

We posit that the BAS concept and resulting MESC provide a process that can be used for every project that includes complex science and technology, and which requires the participation of multi-disciplinary groups.

Finally, there have been claims and counter claims that a scientific issue relating and/or pertinent to medicine, environment, nutrition, engineering, etc. has been "settled", beyond any further discourse or debate. According to the BAS concept and resulting MESC, a scientific claim is settled only if (and only if) it meets the *Reproducibility Principle* such that:

> *A scientific claim is settled if and when any investigator who has the necessary skills, and, depending upon the issue, requisite equipment and facilities, is able to reproduce and confirm it.*

BIBLIOGRAPHY

American Association of Engineering Societies (AAES). Energy and the environment: a general position paper of the American Society of Mechanical Engineers adopted by AAES. Washington, DC: AAES; 1989.

American Bar Association (ABA). Judge Jack's opinion in THE SILICA MDL, Judge Jack uncovers a "phantom epidemic".www.abanet.org/litigation/litigationupdate2007/september_committee-undamaged . Accessed May 2008.

American Chemical Society (ACS). Chemical professional's code of conduct; 2007. www.portal.acs.org. Accessed June 2009.

American Institute of Chemical Engineers (AICHE). Code of Ethics; 2003. www.aiche.org/about/code.aspx. Accessed June 2009.

American Medical Association (AMA). Declaration of professional responsibility; Medicine's social contract with humanity. 2001, www.ama.assn.org/ama/pub/physician-resources /medical-ethis/code-medical-ethics.shtml Accessed June 2009.

American Society of Mechanical Engineers/Institute for Regulatory Science (ASME/RSI). *Assessment of Technologies Supported by the Office of Science and Technology, Department of Energy; Results for Fiscal Year 2002.* New York: ASME; 2002.

American Thoracic Society (ATS). Diagnosis and initial management of nonmalignant diseases related to asbestos. *Am. J. Respir. Crit. Care Med.* 2003; 17:691-715.

Bell MS. *Lavoisier in the Year One; the Birth of a New Science in the Age of Revolution.* New York: Norton and Co.; 2005.

Bellinger DC, et al. Neuropsychological and renal effects of dental amalgam in children. *J. Am. Med. Assoc.* 2006;295: 1775-1783.

Centers for Disease Control and Prevention (CDC). Ethical consideration for B-readers. www.cdc.gov/niosh/topics/chestradiography/breader-ethics. Accessed May 2008.

Daubert v. Merrell Dow. 509 U.S.579; 1993.

Department of Health and Human Services (DHHS). Protection of human subjects. Code of Federal Regulations; 45 CFR Part 46. Washington, DC: Government Printing Office.

Doll R, Peto R. The causes of cancer: Quantitative estimates of avoidable risks of cancer in the United States today. *J. Nat. Cancer Inst.* 1981; 66: 1192-1308:.

Einstein A. Science and religion, in *Out of My Later Years*. New York: NY Philosophical Library; 1950.

Endangered Species Act. 35 U. S. C. 1531 et seq.

Federal Water Pollution Control Act. 33 U. S. C. 1251 et seq.

Energy Policy Act of 2005. 42 U. S. C. 13381et seq.

Gai M, et al. Upper limits ion neutron and big gamma-ray emission from cold fusions. *Nature* 1989; 340: 29-34.

Giordano J. Good as gold? The randomized controlled trial: pragmatic and ethical issues in pain research. *Am J. Pain Management* 2006; 16:68-71.

Giordano J, Boswell MV. Evidence-based or evidence-biased: on the need to consider stakeholder values in research assessment. *Pain Physician* 2009; 12:9-12.

Giordano J, Gordijn B. (eds) *Scientific and Philosophical Perspectives in Neuroethics*. Cambridge: Cambridge University Press; 2010.

Gitlin JN, et al. Comparison of "B" readers interpretation of chest radiographs for asbestos related changes. *Acad. Radiol.* 2004;11: 843-856.

Gofman JW. *Radiation and Human Health*. San Francisco: Sierra Club Books; 1981.

Gould SJ. *Rock of Ages: Science and Religion in the Fullness of Life*. New York: The Ballantine Publishing Group; 1999.

Hahn O, Strassmann F. Über den Nachweis und das Verhalten der bei Bestrahlung des Urans mittels Neutronen entstehenden Erdalkalien (On the detection and characteristics of the alkaline earth metals formed by the irradiation of uranium with neutrons). *Naturwissenschaften* 1939; 27:11-15.

Helmholtz H. von, Die Lehre von den Tonempfindungen als physiologische Grundlage für die Theorie der Musik (On the sensation of tone as the physiological basis for the theory of music) 1863.

Higgins W. *Comparative Law of the Phlogiston and Antiphlogiston Theories with Inductions to Which Is Annexed an Analysis of the Human Calculus with Observations on Its Origin, Etc.* London: J. Murray; 1789.

Hill AB. The environment and disease: Association or causation. *Proc. Roy. Soc. Med.* 1965; 58:295-300.

Höfer T. Przyrembel, H.; Verleger, S. New evidence for the theory of the stork. *Pediatr. Prenat. Epidemiog.* 2004; 18: 88-92.

International Agency for Research of Cancer (IARC). *Polycyclic Aromatic Hydrocarbons. Polynuclear Aromatic Compounds, Part 3, Industrial Exposure in Aluminium Production, Coal Gasification, Coke Production, and Iron and Steel Founding. Vol. 35*; 1998. Geneva, Switzerland WHO Press. http://monographs.iarc.fr. Accessed June 2008.

Kantrowitz A. Proposal for an institution for scientific judgment. *Science* 1967; 156:763.

Kronick DA. Peer review in the 18th century scientific journalism. In: *Peer Review in Scientific Publishing.* Chicago, IL: Council of Biological Editors; 1991.

Kumho Tire v. Carmichael. 526 U.S.137; 1999.

Mandel R. *A Century of Peer Review.* Bethesda, MD: National Institutes of Health; 1996.

Munnangi S, Sonnenberg A. Time trends of physician visits and treatment patterns of peptic ulcer disease in the United States. *Arch. Intern med.* 1997; 175: 1489-1494.

National Institutes of Health (NIH). *Helicobacter Pylori* in peptic ulcer disease. National Institutes of Health consensus development panel on *Helicobacter Pylori* in peptic ulcer disease. *J. Am. Med. Assoc.* 1994; 272:65-69.

National Institutes of Health (NIH). *The Hippocratic Oath.* www.nlm.nih.gov/hmd/greek/greek_oath. Accessed June 2009.

National Research Council (NRC). *The Effects on Populations of Exposure to Low Levels of Ionizing Radiation.* (BEIR III Report). Washington, DC: National Academy Press; 1980.

National Research Council (NRC). *Agriculture, Forestry, and Fishing Research at NIOSH.* Washington, DC: National Academy Press; 2008a .

National Research Council (NRC). *Conflict of Interest in Medical Research, Education, and Practice.* Washington, DC: National Academy Press; 2008b.

National Research Council (NRC). *Ballistic Imaging.* Washington, DC: National Academy Press; 2008c.

National Research Council (NRC). *Assessment of Corrosion Education.* Washington, DC: National Academy Press; 2009.

Office of Management and Budget (OMB). Final information quality bulletin for peer review. *Fed. Reg.* 2005; 70: 2664-2677.

Parker KP, DeWees MA, Radelet ML. Race, death penalty, and wrongful convictions. *Crim. Justice, Mag.* 2003;18: 1-7. www.abanet.org/crimjst/spring2003 /death_penalty. Accessed in June 2008.

Sheldon D, et al. *Wetlands in Washington State. Volume 1: A Synthesis of Science.* Ecology publication 05-06-006. (March) 2006. www.ecy.wa.gov/pubs.0506006. Accessed March 2008

Sullivan PJ, et al. Defining and implementing best available science for fisheries and environmental science, policy and management. *Fisheries* 2006; 31; 460-465.

Uscinski RH, McBride DK. The shaken baby syndrome: origins and further hypotheses. *Neuro. Med. Chir (Tokyo)* 2008; 48: 151-156.

Vienna NJ, Polan AK. Incidence of low birth weight among Love Canal residents. *Science* 1984; 226: 1217-12-19.

Warren JR, Marshall BJ. Unidentified curved bacilli in the stomach of patients with gastritis peptic ulceration. *Lancet* 1984; 1(8390): 1311-1315.

Williams DE, et al. Upper bounds on cold on cold on cold fusion in electrolytic cells. *Nature* 1989; 342; 375-384.

United States Code (USC). Solid Waste Disposal Act U.S.C.6901 et seq.;1985.

APPENDIX I: BAS Terminology

The terminology used in the BAS concept is also used elsewhere with conflicting and sometimes unclear definitions. The following definitions are intended to provide a reasonably precise description of terminologies used in BAS.

Assessment: Critical evaluation of a subject.

Assessment Panel: A group of individuals chosen to perform scientific assessments.

Conflict of Interest: A circumstance that would lead an individual having to choose between scientific judgment and other interests, including personal interests.

Consensus Panel: A group of individuals chosen to reconcile conflicting information—particularly in Evolving Science and Borderline Science.

Consensus Process: A formal process that attempts to reconcile inconsistent or contradictory scientific information.

Independence: Lack of conflict of interest.

Independent Consensus Process: A formal process that attempts to reconcile inconsistent or contradictory scientific information using independent peers.

Independent Peer Review: Critical evaluation of a project by independent peers.

Independent Scientific Assessment: Critical evaluation of a scientific subject by independent peers.

Peer: An individual who is capable of performing the project or that segment of the project that is being reviewed, with little or no study.

Peer Review: Critical evaluation of a project by peers.

Primary Sciences: Sometimes called natural sciences, primary sciences consist of physics; chemistry; biology; various applications of these scientific disciplines including medical sciences; mathematics; the entire field of engineering; and economics.

Project: A study; a program; a report, an article; an activity; or any other technical subject that is planned, is in progress, or is completed.

Review Panel: A group of individuals chosen to perform peer review.

Scientific Assessment: Critical evaluation of a scientific subject by independent peers.

Social Sciences: Those branches of science that is not included in the definition of primary sciences.

Sponsor: An entity seeking peer review or scientific assessment including governmental agencies at all levels, industry, academia, and nonprofit organizations.

Subject: A scientific, programmatic, or other areas requiring study or assessment.

Technical Assessment: Evaluation of a scientific subject by individuals who are not necessarily peers or independent.

Technical Advice: An evaluation of a project by one or more individuals who are not necessarily peers or independent.

Technical Review: Virtually the same as Technical advice consisting of a review of a project by individuals who are not necessarily peers or independent.

APPENDIX II: List of Acronyms

AP	Assessment Panel
BAI	Best Available Information
BAS	Best Available Science
BASAP	Best Available Science Assessment Panel
B-Reader	An expert who evaluates the radiographic evidence for occurrence of asbestosis
CDC	Centers for Disease Control and Prevention
CP	Consensus Panel
EPA	U.S. Environmental Protection Agency
HTL	Hypothesis, Theory, Law: the conventional scientific evolution
IRB	Institutional Review Board
MESC	Metrics for Evaluation of Scientific Claims

NRC	National Research Council: the research arm of the National Academy of Sciences, National Academy of Engineering, and the Institute of Medicine
OMB	Office of Management and Budget: a U.S. Agency affiliated with the White House
RP	Review Panel
SWDA	Solid Waste Disposal Act
USP	Universal Scientific Principles: principles used by virtually all scientific disciplines

APPENDIX III: Author Biographies

Betty R. Love

Betty R. Love is Executive Vice President of the Institute for Regulatory Science (RSI). In that capacity, she is responsible for the management of the day-to-day operations at RSI, and administratively manages several projects including the program dealing with peer review and scientific assessment. She has designed and implemented a comprehensive and unique Reviews Database that contains more than 1,000 potential candidates. Her research has centered on the stakeholder participation process, including a unique approach on the notification and treatment of various categories of stakeholders.

She has served in a number of progressively responsible positions throughout her professional life as Director of Personnel at the National Alliance of Business in Washington, D.C. She was also Personnel Officer in the Employee Relations Department of the University of Maryland at Baltimore, where she was directly responsible for the design, development, and installation of a database system to track faculty and staff evaluations, suspensions, physical evaluations, and employee recognition programs. While at Temple University in Philadelphia, she was not only the Director of the Department of Information and Training, but also deputy to the Associate Vice President for the Office of Environmental Health and Safety (EHS). As the administrative manager of EHS, she assumed the overall responsibility for the management, coordination, and supervision of EHS's daily operations.

As the University's Director of Information and Training, she was responsible for the development, coordination, and implementation of a significant number of training courses in a broad range of areas such as radiation safety; chemical hygiene; hazard communication; bloodborne and airborne pathogens; fire safety; and emergency management. She developed a state-of-the art training database to identify, schedule, and track those employees who required federal, state, and legally-mandated training. She was also responsible for the organization of a number of workshops and conferences, including the International Dixy Lee Ray Conference, held annually for several years.

Betty Love has served as the Assistant Editor and Managing Editor of *Technology*, a journal that traces its roots to the *Journal of the Franklin Institute*, one of the oldest technical journals in the U.S. In addition, she is responsible for the management of an extensive publications program. She has published a number of peer-reviewed papers, reports, and other publications, and is the primary author of the *Manual for Public and Stakeholder Participation*.

Betty Love was educated at Virginia State University, Petersburg, Virginia, where she received a B.S. degree in Business Administration/Education. She received an M.S. degree in Developmental Clinical Psychology from Antioch University, Yellow Springs, Ohio. She also pursued additional education at the University of Baltimore School of Law, and completed course work in computer science at the Johns Hopkins University.

A. Alan Moghissi

A. Alan Moghissi is currently the President of the Institute for Regulatory Science (RSI), a non-profit organization dedicated to the idea that societal decisions must be based on Best Available Science (BAS). The activities of the Institute include research, peer review of scientific projects, scientific assessments, and science education at all levels—particularly education for minorities. Previously, Alan Moghissi was Associate Vice President for Environmental Health and Safety at Temple University in Philadelphia, PA and Assistant Vice President for Environmental Health and Safety at the University of Maryland at Baltimore. As a charter member of the U.S.

Environmental Protection Agency, he served in a number of capacities, including Principal Science Advisor for Radiation and Hazardous Materials; and Manager of the Health and Environmental Risk Analysis Program. Alan Moghissi's research has dealt with diverse subjects ranging from measurement of pollutants to biological effects of environmental agents. A major segment of his research has been on the development of the BAS concept and its application to the scientific foundation of laws, regulations, judicial decisions, and public information. He has been affiliated with a number of universities. He was a visiting professor at Georgia Tech and the University of Virginia, and was also affiliated with the University of Nevada and the Catholic University of America.

He has published in excess of 400 papers, reports, books, and other materials. In addition, he led the preparation of about 300 reports that deal with peer review of government activities at federal, state, and local levels. He has also authored reports on peer reviews performed for the U.S. Congress and scientific assessments responding to the needs of various organizations at the national and international level. Alan Moghissi was the Editor-in-Chief of *Environment International* and *Waste Management*. He was also Editor-in-Chief of *Technology*, which traces its roots to the Journal of the Franklin Institute, one of America's oldest technical journals. Alan Moghissi is an honorary member of the National Council on Radiation Protection and Measurements; a member of the International Academy of Indoor Air Sciences; an Academic Councilor of the Russian Academy of Engineering; an Academic Councilor of Universidad Popular Autonoma del Estado Puebla, a Mexican University; and a Senior Fellow and Member of the Board of Regents at the Potomac Institute for Policy Studies. He has testified numerous times to Committees of both the U.S. Senate and House of Representatives; has served on a number of national and international panels; and is the recipient of numerous awards, including the Distinguished Career Award of the EPA. He was appointed by Secretary of State as a Commissioner of the U.S. Commission on UNESCO. Alan Moghissi received his education at the University of Zurich, Switzerland, and Technical University of Karlsruhe in Germany, where he received a doctorate degree in physical chemistry.

Sorin R. Straja

Sorin Straja is Vice President for Science and Technology at the Institute for Regulatory Science (RSI). In that capacity, he has served as the primary Technical Secretary of all peer reviews, scientific assessments, and related activities. A key responsibility of Technical Secretary is to extract information from a variety of sources to prepare "Project Summaries" for inclusion in peer review and scientific assessment reports.

Previous to his current position, Sorin Straja was Director of the Department of Occupational Safety and Health and a Professor of Biostatistics and Epidemiology in the School of Dentistry, both at Temple University. Prior to that, he worked at the University of Maryland at Baltimore, a medical and research university, where he performed modeling studies in the areas of occupational safety, movement of pollutants in the environment (notably in the food chain), and applied research on safety issues.

Sorin Straja has been involved in broad areas of science and engineering ranging from unit operations of chemical engineering to risk assessment and microbiology. His particular interest is in statistical assessment, modeling, and related activities. Sorin Straja has been able to evaluate voluminous documents and extract relevant information for the intended purpose as demonstrated by the fact that he has prepared Project Summaries for over 300 peer reviews and scientific assessments with topics covering chemistry, physics, medicine, and several engineering disciplines.

Sorin Straja has more than 40 publications, including two books. He was an editor of *Environment International*, and a contributing editor to *Technology*, with responsibility for extracting relevant features from reports on scientific consensus for publication. Sorin Straja received an M.S. degree in Industrial Chemistry and Ph.D. degree in Chemical Engineering from Polytechnic Institute of Bucharest.

Michael S. Swetnam

Michael S. Swetnam assisted in founding the Potomac Institute for Policy Studies in 1994. Since its inception, he has served as Chairman of the Board and currently serves as the Institute's Chief Executive Officer. Prior to forming the Potomac Institute for Policy Studies, Michael Swetnam was Vice President of Engineering at the Pacific-Sierra Research Corporation, Director of Information Processing Systems at GTE, and Manager of Strategic Planning for GTE Government Systems. Prior to joining GTE, he worked for the Director of Central Intelligence as a Program Monitor on the Intelligence Community Staff. In that capacity he was responsible for the development and presentation to Congress of the budget of the National Security Agency, and helped develop, monitor, and present to Congress the DOE Intelligence Budget. He was also assigned as the Intelligence Committee staff representative to intergovernmental groups that developed the INF and START treaties. He assisted in presenting these treaties to Congress for ratification. His collateral duties included serving as the host to the Director of Central Intelligence's Nuclear Intelligence Panel and Co-Chairman of the Science & Technology Requirements Analysis Working Group. From 1990 to 1992, Michael Swetnam served as a Special Consultant to President Bush's Foreign Intelligence Advisory Board (PFIAB) where he provided expert advice on Intelligence Community issues including budget, community architecture, and major programs. He also assisted in authoring the Board's assessment of Intelligence Community support to Desert Storm/Shield. He also served in the U.S. Navy for 24 years as an active duty and reserve officer, Special Duty Cryptology.

Michael Swetnam has served and continues to serve on numerous panels and committees. He is currently a member of the Technical Advisory Group to the United States Senate Special Select Committee on Intelligence. In this capacity, he provides expert advice to the U.S. Senate on the R&D investment strategy of the U.S. Intelligence Community. He also served on the Defense Science Board (DSB) task force on Countering Terrorism, and the task force on Intelligence support to the war on terrorism. He has served in several public and community positions including Northern United Kingdom Scout Master (1984-85); Chairman, Term Limits Referendum Committee (1992-93); President (1993) of the Montgomery County Corporate Volunteer Council, Montgomery County Corporate Partnership for Managerial Excellence (1993); and the Maryland Business Roundtable (1993). He is also on the Board of Directors of Space and Defense Systems Inc., Dragon Hawk Entertainment Inc., and the governing board of The Potomac Institute of New Zealand.

Michael Swetnam has authored and edited several books and articles including: "Cyber Terrorism and Information Warfare," a four volume set he co-edited; "Usama bin Laden's al-Qaida: Profile of a Terrorist Network," co-authored with Yonah Alexander; and "ETA: Profile of a Terrorist Group," co-authored with Yonah Alexander and Herbert M. Levine. He received a BS degree in Electrical Engineering from University of Oklahoma.

INDEX

A

B

C

D

E

F

G

H

I

J

K

L

M

N

O

P

Q

R

S

T

U

V

W